# ARCHITECTURE EXPLORATION FOR EMBEDDED PROCESSORS WITH LISA

# Architecture Exploration for Embedded Processors with LISA

by

**Andreas Hoffmann**
*LISA Tek GmbH, Germany*

**Heinrich Meyr**
*Aachen University of Technology, Germany*

and

**Rainer Leupers**
*Aachen University of Technology, Germany*

KLUWER ACADEMIC PUBLISHERS
BOSTON / DORDRECHT / LONDON

A C.I.P. Catalogue record for this book is available from the Library of Congress.

ISBN 978-1-4419-5334-6

---

Published by Kluwer Academic Publishers,
P.O. Box 17, 3300 AA Dordrecht, The Netherlands.

Sold and distributed in North, Central and South America
by Kluwer Academic Publishers,
101 Philip Drive, Norwell, MA 02061, U.S.A.

In all other countries, sold and distributed
by Kluwer Academic Publishers,
P.O. Box 322, 3300 AH Dordrecht, The Netherlands.

*Printed on acid-free paper*

*Dedicated to my wife Sabine,*
*my daughter Pauline, and*
*my parents Erich and Sigrid.*

# Contents

# Foreword

Already today more than 90% of all programmable processors are employed in embedded systems. This number is actually not surprising, contemplating that in a typical home you might find one or two PCs equipped with high-performance standard processors, but probably dozens of embedded systems, including electronic entertainment, household, and telecom devices, each of them equipped with one or more embedded processors. Moreover, the electronic components of upper-class cars incorporate easily over one hundred processors. Hence, efficient embedded processor design is certainly an area worth looking at.

The question arises why programmable processors are so popular in embedded system design. The answer lies in the fact that they help to narrow the gap between chip capacity and designer productivity. Embedded processors cores are nothing but one step further towards improved design reuse, just along the lines of standard cells in logic synthesis and macrocells in RTL synthesis in earlier times of IC design. Additionally, programmable processors permit to migrate functionality from hardware to software, resulting in an even improved reuse factor as well as greatly increased flexibility.

The need for very high efficiency in embedded systems poses new constraints on processor design. While an off-the-shelf general-purpose processor might be able to meet the performance requirements of many embedded applications, it is certainly inefficient for a given application, particularly when it comes to portable, battery-driven consumer electronics or telecom devices. As a consequence, in the past one or two decades we have seen a growing variety of domain specific processors, like ultra low cost/low power microcontrollers or DSPs for highly efficient signal processing. More recently, dedicated processors for networking applications (NPUs) are receiving growing interest and market shares, too. Application-specific instruction-set processors (ASIPs), frequently designed in-house by system providers for the sake of better product differentiation, IP protection, or royalty cost reduction, are yet another form of the desired compromise between flexibility and computational efficiency.

Embedded processors, particulary ASIPs, must often be developed by relatively small teams and within stringent time constraints. Hence, automation of the processor design process is clearly a very important issue. Once an HDL model of a new processor is available, existing hardware synthesis tools enable the path to silicon implementation. However, embedded processor design typically begins at a much higher abstraction level, even far beyond an "instruction-set", and involves several architecture exploration cycles before the optimum hardware/software match has been found. In turn, this exploration requires a number of tools for software development and profiling. These are normally written manually – a major source of cost and inefficiency in processor design so far.

The LISA processor design platform (LPDP) presented in this book addresses these issues and results in highly satisfactory solutions. Work on the LISA language and tooling was initially inspired just by the need for fast instruction-set simulation. However, it was early on recognized by Andreas Hoffmann that the initial goal could be far exceeded towards a complete and practical processor design system. The LPDP covers all major high-level phases of embedded processor design and is capable of automatically generating almost all required software development tools from processor models in the LISA language. It supports a profiling-based, stepwise refinement of processor models down to cycle-accurate and even RTL synthesis models. Moreover, it elegantly avoids model inconsistencies otherwise omnipresent in traditional design flows.

The next step in design reuse is already in sight: SoC platforms, i.e., partially predesigned multi-processor templates that can be quickly tuned towards given applications thereby guaranteeing a high degree of hardware/software reuse in system-level design. Consequently, the LPDP approach goes even beyond processor architecture design. The LPDP solution explicitly addresses SoC integration issues by offering comfortable APIs for external simulation environments as well as clever solutions for the problem of both efficient and user-friendly heterogeneous multiprocessor debugging.

The work presented in this book is certainly not the first approach towards automating embedded processor design. However, it is definitely among the most complete, advanced, and flexible ones, based on a huge amount of knowledge on hardware and software design processes, as well as continuous tight interaction with designers in academia and industry "torturing" the LPDP in day-by-day use for real-life projects. We believe, the results are appealing from both a scientific and a user perspective. We hope the reader will be convinced after having read this book!

October 2002 H. Meyr, R. Leupers

# Preface

This book documents more than five years of research carried out at the Institute for Integrated Signal Processing Systems (ISS) at the Aachen University of Technology (RWTH Aachen). It mirrors the results of my PhD thesis which was submitted to the faculty of electrical engineering and information technology (FB6) on May 15th, 2002, with the title "A Methodology for the Efficient Design of Application-Specific Instruction-Set Processors Using the Machine Description Language LISA".

The research on the machine description language LISA and the associated tooling was originally motivated by the tedious and error prone task to write instruction-set simulators manually. At this time (1996), the ISS already had deep knowledge in instruction-set simulation technology, especially in the field of compiled instruction-set simulation. Obviously, the design of the language was strongly influenced by the requirements of automatic simulator generation. In the following years, the language was enhanced to be capable of retargeting code generation tools like assembler and linker, profiling tools, and the system integration of the generated instruction-set simulators. Besides the software development tools also the generation of an implementation model of the target architecture is supported in the latest version of LISA. By this, the complete processor *design* is enabled instead of just automating the realization of software development tools. One of the primary goals when designing the language and the tooling was to cover a wide range of architectures – from digital signal processors (DSPs) to micro-controllers ($\mu$Cs) and special purpose architectures.

All in all, more than 40 man-years of effort by PhD students have been spent to develop both the language and the associated tooling. This does not include the work of numerous part time student workers, semester works, and master theses. A special thanks to my PhD colleges at the ISS working in the LISA team without whose engagement the results presented within the scope of this book would have never been achievable. Especially my former master students Achim Nohl, Gunnar Braun, and Oliver Schliebusch have made major contributions to this book. Furthermore, I want to mention Tim Kogel, Andreas Wieferink, and Oliver Wahlen which I am thankful for reviewing the manuscript and for their contributions in the area of system integration and retargetable compilation. Besides, I am grateful that my company LISATek, who commercialized the LISA technology in the beginning of 2001, granted me time to work on this book. A special thanks to my two advisers and co-authors of this book, Prof. Heinrich Meyr and Prof. Rainer Leupers. Moreover, I want to thank my wife and my family for their patience when writing this book.

The technology and the LISA language presented in this book are just a snap-shot of the research work carried out at the ISS. Today, there are still 10 PhD students working on LISA related topics – one major new area to investigate is retargetable compilation based on LISA. Furthermore, in the area of HDL-code generation there is an endless amount of interesting research work to do. Consequently, readers of this book interested in the current status of the technology should consider looking at the the webpages "http://www.iss.rwth-aachen.de/lisa" for the status of the research work and "http://www.lisatek.com/products" for the status of the productization of the research work at LISATek Inc.

ANDREAS HOFFMANN

Chapter 1

# INTRODUCTION

In consumer electronics and telecommunications high product volumes are increasingly going along with short life-times. Driven by the advances in semiconductor technology combined with the need for new applications like digital TV and wireless broadband communications, the amount of system functionality realized on a single chip is growing enormously. Higher integration and thus increasing miniaturization have led to a shift from using distributed hardware components towards heterogeneous system-on-chip (SOC) designs [1]. Due to the complexity introduced by such SOC designs and time-to-market constraints, the designer's productivity has become the vital factor for successful products. Today, development time is often more valuable than MOPS (million operations per second) and the ratio is rising (see figure 1.1). For this reason a growing amount of system functions and signal processing algorithms is implemented in software rather than in hardware by employing embedded processor cores. The programmability helps to raise the designer's productivity and the flexibility of software allows late design changes and provides a high grade of reusability, thus shortening the design cycles.

As a consequence, embedded processors, especially digital signal processors (DSPs), have hit the critical mass for high-volume applications [2]. According to market analysts, the market of embedded processors is growing much faster than the market for information technology in general. Many segments of the embedded processor market are *consumer markets*, with very short product lifetimes and short market windows. Today, the entire digital wireless industry operates e.g. with DSP-enabled handsets and base stations. The mass-storage industry depends on embedded processors to produce hard-disk drives and digital versatile disc players. Ever-increasing numbers of digital subscriber line and cable modems, line cards, and other wired telecommunications equipments are based on programmable components. Hearing aids, motor control, consumer audio gear such as internet audio are just some of the many mass market applications in which embedded processors are frequently found today. More specialized applications include image processing, medical instrumentation, navigation, and guidance. Besides, using embedded processors in devices that previously relied on analog circuitry – such as digital cameras, digital camcorders, digital personal recorders, internet radios, and internet telephones – provides revolutionary performance and functionality that could never have been achieved by merely improving analog designs. The increasing availability of system level integration (SLI) heralds a vast array of even more innovative products.

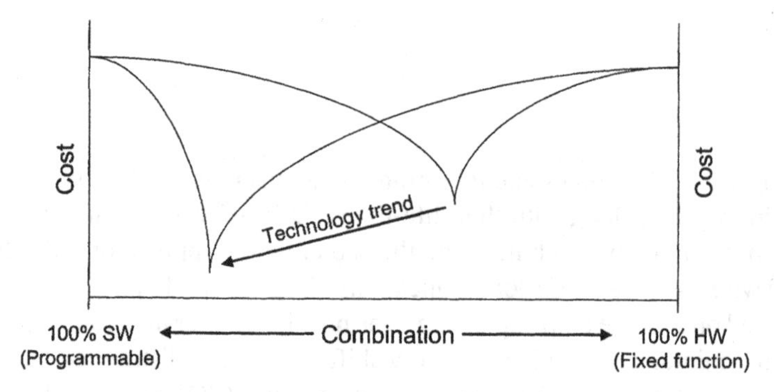


*Figure 1.1.* Combining software and hardware for the lowest cost system design[1].

In the next decade, the continued growth of embedded processors will mainly depend on developments in three key areas of technology [3, 4]:

- the underlying manufacturing processes,
- the embedded core and chip architectures, and
- the software for development and applications.

An additional factor and obviously the most difficult to anticipate is innovation. In only a few years from now, designers of embedded processors will be dealing with architectures that integrate hundreds of millions of on-chip transistors and deliver performance in the order of trillions of instructions per second [5, 6]. Determining how to use all this processing power goes beyond conventional engineering methods.

In the following, the term *embedded processor* representing a large variety of processors is illuminated in more detail by partitioning processors in different

[1]Cost can be defined in terms of financing, design cost, manufacturing cost, opportunity cost, power dissipation, time-to-market, weight, and size.

processor categories. Furthermore, the advent of one particular flavor of embedded processors is motivated – application-specific instruction-set processors (ASIPs).

## 1. Processor Categories

In the current technical environment, developers design general-purpose embedded systems as off-the-shelf (OTS) parts for reuse in numerous products. As their specific applications are unknown, these designs must incorporate both generality and completeness. Specialization involves departures from both characteristics with respect to structure and functionality. Extreme specialization results in an application-specific system or a processor designed for a single application. In contrast to that, a domain-specific system or processor has been specialized for an application domain, e.g. HDTV or set-top boxes, but not for a specific application.

At the one end of the spectrum, application-specific architectures provide very high performance and high cost performance, but reduced flexibility. At the other end, general-purpose architectures provide much lower performance and cost performance, but with the flexibility and simplicity associated with software programming. For a desired performance level on a specified workload, specialization minimizes the logic complexity and die size – and thus the fabrication cost – of an embedded computer system or processor. Figure 1.2 shows a classification of embedded processors proposed by Marwedel [7]. This *processor cube* results from using three main criteria for classifying processors: availability of domain-specific features, availability of application-specific features, and the form in which the processor is available.

The meaning of the dimensions of this processor cube and their values are as follows:

1 *Form in which the processor is available*
At every point in time, the design and fabrication processes for a certain processor have been completed to a certain extent. The two extremes considered here are represented by completely fabricated, packaged processors and by processors which just exist as a cell in a CAD system. The latter is called a core processor.

2 *Domain-specific features*
Processors can be designed to be domain-specific. Possible domains are digital signal processing or control-dominated applications. DSPs [8, 9] contain special features for signal processing: multiply/accumulate instructions, specialized (heterogenous) register-sets, multiple ALUs, special DSP addressing modes, e.g. ring buffers, and saturating arithmetic operations.

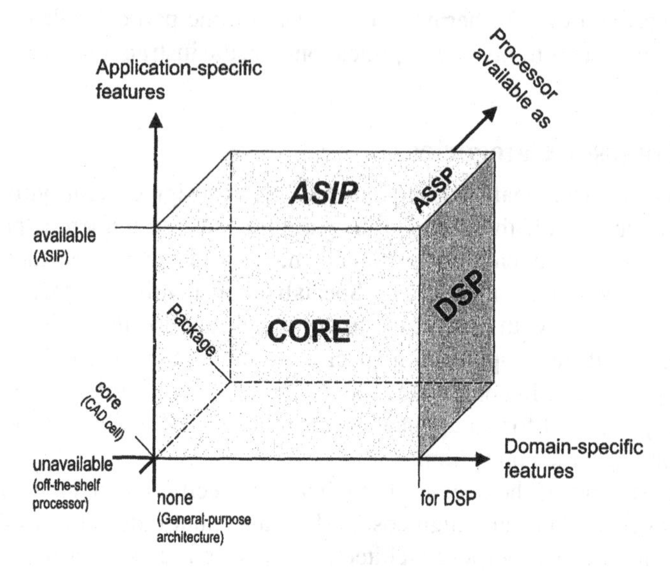


*Figure 1.2.* Cube of processor types.

3 *Application-specific features*
The internal architecture of a processor may either be fixed or still allow configurations to take place. The two extremes considered here are: processors with a completely fixed architecture and ASIPs. Processors with a fixed architecture or off-the-shelf processors have usually been designed to have an extremely efficient layout. ASIPs are processors with an application-specific instruction-set (cf. section 2.2). Depending on the application, certain instructions and hardware features are either implemented or unimplemented. ASIPs have the potential of requiring less area or power than off-the-shelf processors. Hence, they are popular especially for low-power applications. The special class of ASIPs optimized for DSP is also called application-specific signal processors (ASSP). These processors correspond to one of the edges of the processor cube.

In the following it will be motivated why ASIPs are supposed to supersede general-purpose architectures in future SOC designs. For integration into SOC, these architectures have to be available as cores. The illustrations are independent of domain-specific processor features.

## 2. Advent of ASIPs in System-on-Chip Design

With the development of the sturdy, low-priced Model-T in 1908, Henry Ford made his company the biggest in the industry. However, it was not until 1914 that Ford achieved the breakthrough. At this point in time, Henry Ford revolutionized the manufacturing processes by moving to an assembly line that finally enabled Ford to produce far more cars than any other company. The Model-T and mass production made Ford an international celebrity.

The question arises, if such a fundamental change is also a necessity for the area of embedded processors to cope with the requirements of future SOC designs. In the opinion of the authors the answer is: yes. Future SOC will be mainly dominated by embedded memory and a large variety of application-specific instruction-set processors operating on this memory. Stated in another, more striking way: software will replace gates.

## 2.1 Technology Status and Trends

In 1965, when preparing a talk, Gordon Moore of Intel noticed that up to that time microchip capacity had seemed to double each year [10]. With the pace of change having slowed down a bit over the last years, we have seen the definition of Moore's law change to reflect that the doubling occurs only every 18 months. In other words, transistor count growth at a compound rate of 60%. Process and architectural changes act to drive Moore's law for transistor count and also for performance. CPU cycle time measured in megahertz (MHz) tends to track Moore's law, following the same path as transistor count.

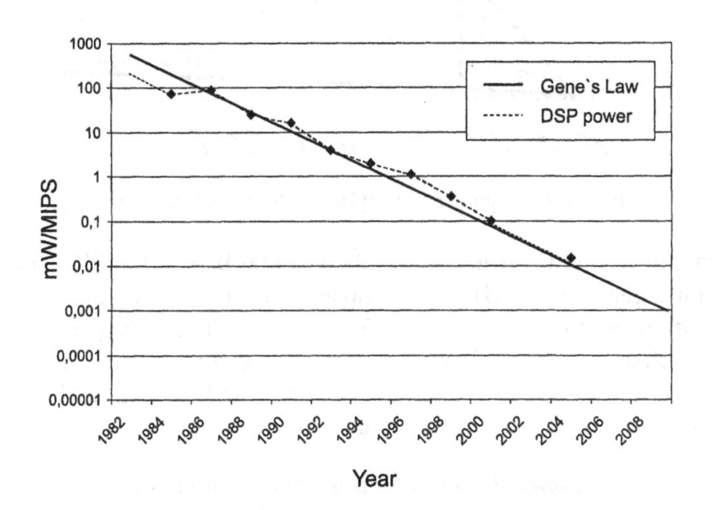


*Figure 1.3.* Power dissipation trends – *Gene's Law.*

The same applies to power dissipation in the opposite way. According to Gene Frantz of Texas Instruments, DSP power dissipation per MIPS halves every 18 months (see figure 1.3, *Gene's law* [5]). Current projections by Texas Instruments are that by 2002, a 5 million transistor DSP that provides 5,000 MIPS will consume 0.1 mW/MIPS. Ten years later, a DSP with 50 million transistors capable of achieving 50,000 MIPS will run on 1 nanowatt (nW) per MIPS.

As the power efficiency is increasing with Moore's law, the question arises, why ASIPs have become so popular in the last couple of years and are supposed to take away large parts of hard-wired logic from future SOC designs [11, 12]. The reason for this is founded in the algorithmic complexity which is depicted qualitatively in figure 1.4[2].

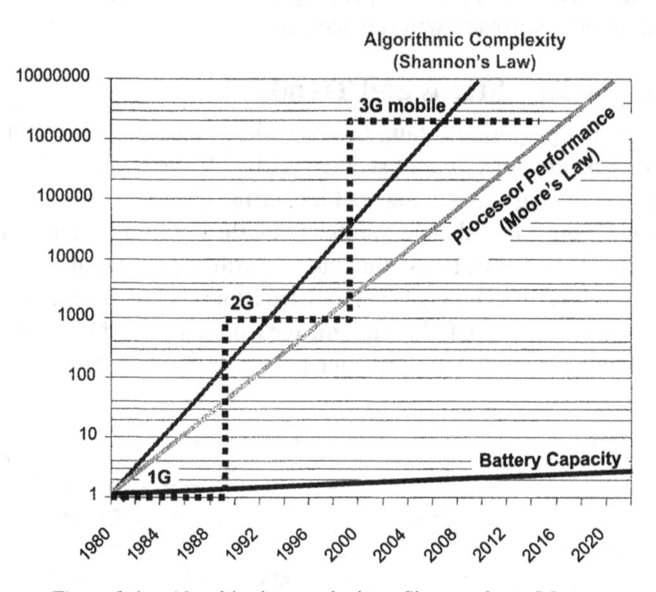


*Figure 1.4.* Algorithmic complexity – Shannon beats Moore.

According to a study carried out by Silicon Valley based MorphICs Technologies, a third generation (3G) mobile phone platform provider, the algorithmic complexity rises stronger than Moore's law. Algorithmic complexity is visualized by Shannon's law [13] which defines the theoretical maximum rate at which error-free digits can be transmitted over a bandwidth-limited channel in the presence of noise. In short it can be stressed that

***Shannon asks for more than Moore can deliver !***

As the battery capacity remains nearly constant, it becomes obvious that new architectural solutions are required to cope with the ever increasing algorithmic complexity introduced by new mobile standards like 3G or even 4G (third/fourth generation mobile phone).

[2]Courtesy of Dr. Ravi Subramanian of MorphICs Technologies Inc.

Using embedded processors like DSPs, the above mentioned gap can only be overcome by employing more processors in the system. However, power consumption of the overall system is extremely critical – not only in mobile applications [14]. The trade-off between energy efficiency (quantified in MOPS/mW) and flexibility of the architecture is depicted in figure 1.5[3].

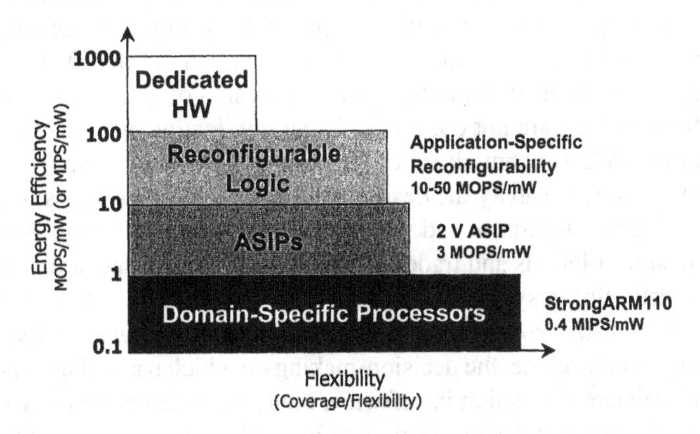


*Figure 1.5.* The energy flexibility gap.

Obviously, the two extremes are on the one hand dedicated, hard-wired logic – commonly known as application-specific integrated circuit (ASIC) – and on the other hand domain-specific embedded processors like the Texas Instruments TMS320C62x for DSP [15] and the ARM7 [16] as a representative for RISC micro-controllers. Dedicated hardware provides the highest energy efficiency while lacking flexibility to adopt to changing applications. Domain-specific embedded processors are very flexible to run a large mix of applications while being less energy-efficient. The gap between full flexibility and hardwired logic is filled by reconfigurable logic and ASIPs. The former is addressed by FPGA (field programmable gate arrays) vendors like XILINX or ALTERA [17, 18] and provides fine-grain reconfigurability while being not very cost-efficient in mass-market applications. The latter is a cost-efficient design paradigm which can be used by system-houses like Nokia or Cisco to design application-specific processors in-house (in case a suited design methodology is available) superseding the need to pay expensive royalties to semiconductor vendors.

## 2.2 Application-Specific Instruction-Set Processors

Application-specific instruction-set processors are designed to address the challenges of application areas for processor-based systems which have diverse

[3]Courtesy of Prof. Jan Rabaey of the University of California, Berkeley.

and often conflicting requirements on the processors at the core of these systems. These requirements can be low power consumption, performance in a given application domain, guaranteed response time, code size, and overall system cost [19].

To meet these criteria, processor characteristics can be customized to match the application profile. Customization of a processor for a specific application holds the system cost down, which is particularly important for embedded consumer products manufactured in high volume. The trade-off involved in designing ASIPs differs considerably from the design of general-purpose processors, because they are not optimized for an *application mix* (such as represented in the SPEC benchmark suite [20]) but target a specific problem class. The primary aim is to satisfy the design goals for that problem class, with performance on general-purpose code being less important. In designing ASIPs, different design solutions and trade-offs between hardware and software components constituting a system have to be explored. This is often referred to as *hardware-software co-design* [21, 22, 23]. Besides hardware-software co-design on system level, i.e. the decision-making on which parts of the system to realize in hardware and which in software, the term is also commonly used in the context of processor design. Here, it refers to the step of tailoring the architecture to the application. Hardware-software co-design of application-specific processors identifies functions which should be implemented in hardware and in software to achieve the requirements of the application, such as code size, cycle-count, power consumption and operating frequency. Functions implemented in hardware are incorporated in an application-specific processor either as new instructions and processor capabilities, or in the form of special functional units. These special functional units may then either be integrated on the chip or implemented as peripheral devices.

According to a survey carried out by Pierre Paulin at STMicroelectronics, already in 1997 ASIPs represented more than two-thirds of the volume of chips used in embedded applications [24]. The forecast indicates that the total percentage is going to rise in future designs. However, the growth rate of ASIPs in SOC designs is strongly dependent on electronic design automation (EDA) to help designing and using those processors more efficiently and reduce the risk. Currently, the lack of a suitable design technology remains a significant obstacle in the development of such systems (cf. chapters 2 and 4).

A new methodology to design and program such ASIPs is presented within the scope of this book. The LISA processor design platform is an environment supporting the automation of embedded processor design. This comprises the support in the architecture exploration phase to map the target application most efficiently to an instruction-set and a micro-architecture, the generation of a synthesizable implementation model, the generation of application software development tools like assembler, linker, simulator, and debugger and the capability to integrate the processor into the system. The presented approach is based on processor models specified in the machine description language LISA [25].

## 3. Organization of this Book

This book is organized as follows: chapter 2 illustrates the traditional design process of embedded processors. From the explanations it becomes obvious that without automation and a unified approach the hurdle to be overcome by system houses to design ASIPs themselves will be too high for many companies. As automation in ASIP design has been addressed by academia and industry in the past, the chapter also reviews existing approaches and closes with the motivation of the work covered within this book.

Chapter 3 works out different processor models which are captured within the LISA language underlying this work. These processor models are required to address different aspects of the processor design: retargeting of software development tools, generation of hardware implementation model, and system integration and verification. Besides, the capability of the language to abstract from the processor architecture on various levels in the domain of architecture and time is shown. This capability is of key importance for a seamless mapping of the application onto a processor architecture.

In chapter 4, the LISA processor design platform (LPDP) is introduced, which has been developed in the last five years at the Institute for Integrated Signal Processing Systems (ISS) at Aachen University of Technology (RWTH Aachen) and is now commercialized at LISATek Inc. LPDP comprises the automatic generation of software development tools for architecture exploration, hardware implementation, software development tools for application design, and co-simulation interfaces from one sole specification of the target architecture in the LISA language.

The following chapters focus on particular phases in the processor design process in more detail. Chapter 5 shows the general work-flow in the design process of application-specific instruction-set processors and presents LISA code segments and tools provided by LPDP supporting this. In a case study, a sample exploration is carried out for a CORDIC angle calculation and a simple ASIP efficiently executing this algorithm is developed.

Chapter 6 deals with the next phase in ASIP design in which the micro-architecture needs to be specified in a hardware description language like VHDL or Verilog and taken through synthesis. Here, it will be shown that large parts of the LISA model resulting from the architecture exploration phase can be reused to generate those parts of the implementation model that are tedious and error-prone to write, but have minor influence on the overall performance of the target architecture – structure, decoder, and controller. A case study carried out with Infineon Technologies on an ASIP for digital video broadcast terrestrial (DVB-T) will prove the presented concept.

Following the phases primarily addressing the architecture design, chapter 7 focuses on software development tools required to program the architecture. Particular emphasis is on different simulation techniques offered by LPDP, which have varying strengths and weaknesses depending on the target architecture and application domain. Again, case studies carried out with real-world DSP and micro-controller architectures will show the quality of the tools by comparing them to the tools provided by the respective architecture vendor.

Finally, chapter 8 addresses the issue of system integration and verification of the processor. Early integrability of simulators to gather realistic stimuli from the system context and flexibility of interfaces to couple the simulators to arbitrary system simulation environments is of key importance from the very beginning in the design process. In a case study, the integration of processor simulators into the commercial CoCentric System Studio [26] environment of Synopsys is demonstrated.

The book closes with a summary and an outlook on open issues and interesting future research topics. Appendices A through D present information on abbreviations used within this book, the grammar of the LISA language, a sample LISA model of the ARM7 $\mu$C architecture, and some details on the ICORE architecture used in the case study on HDL-code generation from LISA.

# Chapter 2

# TRADITIONAL ASIP DESIGN METHODOLOGY

In the current technical environment, ASIPs and the necessary software development tools are mostly designed manually, with very little automation [27]. However, this results in a long, labor-intensive process requiring highly skilled engineers with specialized know-how – a very scarce resource. Most of today's processor design is conducted by embedded processor (EP) and integrated circuit (IC) vendors using a variety of development tools from different sources, typically lacking a well-integrated and unified approach. Engineers design the architecture, simulate it in software, design software for the target application, and integrate the architecture into the system. Each step of this process requires its own design tools and is often carried out by a separate team of developers. As a result, design engineers rarely have the tools nor the time to explore architectural alternatives to find the best solution for their target applications.

The design flow of a processor architecture can be roughly partitioned into four different phases, which are depicted in figure 2.1:

- architecture exploration,
- architecture implementation,
- software application design, and
- system integration and verification.

In the *architecture exploration phase*, the target application requiring application-specific processor support is to be mapped most efficiently onto a dedicated processor architecture. In order to enable the exploration of the wide processor design space, a set of software development tools (i.e. high-level language (HLL) compiler, assembler, linker, and simulator) is required early in the design

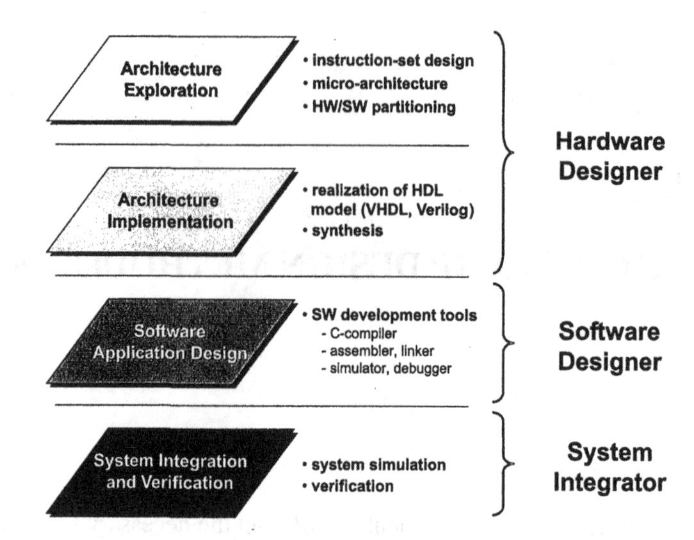


*Figure 2.1.* Four phases in the traditional design process of embedded processors.

process to enable profiling and benchmarking the target application on different architectural alternatives.

In general, the exploration phase is composed of three main tasks, which are closely connected: firstly, the application has to be profiled to determine critical portions that require dedicated hardware support through application-specific instructions. In this context, this task is often referred to as hardware-software partitioning. Secondly, the instruction-set is defined on the basis of the hardware-software partitioning and the application profiling results. Finally, the micro-architecture that implements the instruction-set is fixed.

The process of finding a suitable hardware-software partitioning, instruction-set, and micro-architecture is never performed in a straight top-down fashion but is usually an iterative one that is repeated until a best-fit between the selected architecture and target application is obtained. It is obvious that every change to the architectural specification implies a complete new set of software development tools. As these changes on the tools are carried out mainly manually, this results in a long, tedious, and extremely error-prone task. Besides, the lack of automation makes it very difficult to match the profiling tools to an abstract specification of the target architecture.

In the *architecture implementation phase*, the specified processor has to be transformed into a synthesizable hardware description language (HDL) model. For this purpose, the processor is described in a hardware description language like VHDL [28] or Verilog [29] which can be taken through the standard synthesis flow [30]. With this additional manual transformation it is quite obvious that considerable consistency problems arise between the textual architecture specification, the software development tools, which are usually realized in a HLL like C/C++, and the hardware implementation. The outcome of the architecture implementation phase is a production ready version of the processor architecture, which can be passed to a fab for fabrication.

During the *software application design phase*, software designers require a set of production-quality software development tools to comfortably program the architecture. However, as the demands of the software application designer and the hardware processor designer place different requirements on software development tools, new tools are required. For example, the processor designer needs a cycle-based simulator for hardware-software partitioning and profiling, which is very accurate but inevitably slow, whereas the application designer demands more simulation speed than processor model accuracy. At this point, the complete software development tool-suite is usually re-implemented by hand – consistency problems are self-evident.

Finally, in the *system integration and verification phase*, co-simulation interfaces must be developed to integrate the software simulator for the chosen architecture into a system simulation environment. These interfaces vary with the architecture that is currently under test. Again, manual modification of the interfaces is required with each change of the architecture, which is a tedious and lengthy task.

Besides the above mentioned difficulties in the processor design process caused by different specification languages (textual, C/C++, VHDL, and Verilog), it is frequently the case that separate groups are concerned with the respective design phases within companies developing processor architectures. Both architecture exploration and implementation are usually carried out by the group of *hardware designers*, whereas *software-tool designers* are concerned with the design of production quality software development tools for application design. The system integration and verification is frequently performed by a group called *system designers*, which have in-depth knowledge about the complete system integrating the programmable architecture.

Due to the heterogeneity of people in terms of technical expertise involved in the design of a processor architecture, it is obvious that several problems come up that cause severe inefficiencies in the processor design process. On the one hand, communication is difficult between these groups. Even though it is evident, that communication is often complicated by the fact that the groups are located in different places around the world, it even becomes worse as the groups also speak and think differently. For example, the hardware designer thinks in detailed hardware structures while the software designer sees the architecture only from the instruction-set point of view. On the other hand, information exchange about the processor architecture itself between designers is problematic. Today, information exchange between the groups takes place either on the basis of a textual specification of the instruction-set and the abstracted activities of the micro-architecture or in the form of the implementation model. In the former case, the documentation is frequently faulty and therefore leads to inconsistencies between the processor architecture and the software tools passed to the application designer. In the latter case, the software tools, especially the simulation tools, are realized on the basis of the implementation model. Obviously, the implementation model comprises micro-architectural information that is not needed for the software simulator. However, it is difficult, in most cases even impossible for the software designer to abstract from the implementation model. As a result, the simulation tools are mostly extremely slow and offer an unacceptable simulation speed for the verification of complex applications (cf. chapter 7.4).

In the following, several processor design methodologies introduced by both academia and industry in the last decade are introduced. Many of them only address specific phases of the processor design process or even only specific problems within one phase. The chapter closes with the motivation of this work, which results from the limitations and drawbacks of the related work.

## 1. Related Work

Hardware description languages like VHDL or Verilog are widely used to model and simulate processors, but mainly with the goal to develop hardware. Using these models for architecture exploration and production quality software development tool generation has a number of disadvantages. They cover a huge amount of hardware implementation details which are not needed for performance evaluation, instruction-set simulation (ISS), and software verification. Moreover, the description of detailed hardware structures has a significant impact on simulation speed [31, 32]. Another problem is that the extraction of the instruction-set is a highly complex, manual task and some instruction-set information, e.g. assembly syntax, cannot be obtained from HDL descriptions at all [33].

To overcome this problem, a new class of specification languages was introduced in the last decade which also supports automatic software development tool-kit generation: architecture or machine description languages. There are many publications on machine description languages providing instruction-set models. These languages can be coarsely classified into the following three categories [34, 35]:

- ***Instruction-set centric languages***. The machine description languages nML [36, 37], ISDL [38], Valen-C [39], and CSDL [40] characterize the processor by its instruction-set architecture. In other words, *instruction-set* or *behavior-centric* machine description languages provide the programmer's view of the architecture through the description of the instruction-set.

- ***Architecture-centric languages.*** MIMOLA [41], AIDL [42], and COACH [43] focus on the structural components and connectivity of the processor architecture. The advantage here is that the same description can be used for both synthesis and software tool generation. *Architecture-* or *structure-centric* machine description languages often provide a block-diagram or net-list view of the processor architecture.

- ***Instruction-set and architecture oriented languages.*** Description languages like FlexWare [44], MDes [45], PEAS [46, 47], RADL [48], EXPRESSION [49], and LISA bridge the gap between instruction-set centric languages which are too coarse for cycle-accurate simulation of pipelined architectures and for processor synthesis and hardware description languages which are too detailed for fast processor simulation and for compiler generation. These languages are often referred to as *mixed-level* machine description languages [35].

In the following paragraphs, selected approaches on machine description languages from each of the categories listed above will be briefly introduced. Besides, the limitations with respect to the support of the complete processor design flow are emphasized. In addition to that, other relevant approaches which do not fit into any of the categories from both academia and industry are discussed.

## 1.1 Instruction-Set Centric Languages

Instruction-set centric machine description languages are mostly developed with the prior goal of retargeting HLL compilers. As the architectural information required for this purpose primarily concerns the instruction-set as well as constraints on the sequencing of instructions and instruction latencies, no further information on the micro-architecture is included. This excludes these languages from the retargeting of cycle-based simulators for pipelined architectures or even architecture synthesis.

**The nML Machine Description Formalism.** The language nML [36] was developed at the Technische Universität (TU) Berlin for the description of processor instruction-sets. In nML, the instruction-set is enumerated by an attributed grammar. For software simulation, a rather simple model of execution is proposed: a running machine executes a single thread of instructions, which are held in memory and are addressed by a program counter. The language permits concise, hierarchical processor description in a behavioral style. One of the major shortcomings of nML is the disability to support pipelined instructions and pipeline control mechanisms such as stalls and flushes. Besides the instruction-set simulator SIGH/SIM [50], a code generator called CBC [51] can be generated automatically from machine descriptions in nML.

Linking up with the work of Fauth and Freericks in Berlin, IMEC [52] in Leuven, Belgium, independently developed a code generation tool called CHESS [53] and an instruction-set simulator named CHECKERS [54]. Today, these tools are commercially available from Target Compiler Technologies [55].

Besides TU Berlin and IMEC, the nML formalism was adopted at Cadence and the Indian Institute of Technology. Hartoog of Cadence developed a tool-kit based on nML consisting of interpretive and compiled instruction-set simulator, assembler, disassembler and code generator [56]. Rajesh of Cadence research center India extended nML to the Sim-nML language [57] by constructs to model control-flow and inter-instruction dependencies. By this, pipelines can be modeled as well as mechanisms like branch prediction and hierarchical memories [58]. However, the reported speed of the generated instruction-set simulator (approximately 3,000 insn/sec) is very low. Moreover, even with this extension processors with more complex execution schemes and explicit instruction-level parallelism like the Texas Instruments TMS320C6x cannot be described at all, because of the numerous combinations of instructions.

In many respects LISA incorporates ideas similar to nML. In fact, large parts of the LISA language are following the pattern of nML. This concerns the specification of the instruction-set model in a hierarchical fashion and the structure. Fundamentally different from nML is the way of describing the timing and behavior in the LISA language. Due to the similarity of the two approaches, the above mentioned four characteristics of a processor model – instruction-set, structure, timing, and behavior – are compared more closely in the following.

*Specifying the instruction-set and model hierarchy.* In nML, the description of properties of instructions is captured in operations. One or more operations form an instruction. Operation hierarchy is built up in two ways: by *OR*-rules and by *AND*-rules. An *OR*-rule lists certain alternatives.

**nML** :

```
opn alu = add | sub | or | and
```

lists four possible symbols to derive the non-terminal *alu*. The same construct is known in LISA as a group.

**LISA** :

```
GROUP alu = { add || sub || or || and };
```

In contrast to the *OR*-rule, the *AND*-rule combines several parts. These are listed in the rule's parameter list. The declaration of a parameter consists of an identifying name and a reference to another rule.

**nML** :
```
opn insn( a:src, b:src, c:alu, d:dst )
```

states that the operation *insn* is composed of two derivations of *src*, one of *alu*, and one of *dst*. The same can be described in LISA within the *declare*-section, which announces the operations which are referenced from within the current operation. Both, *AND*- and *OR*-rule, build the operation hierarchy.

Two important attributes of operations describing the instruction-set in nML are *image* and *syntax*. The former is used to describe the binary representation while the latter describes the assembly language mnemonic of the instructions.

**nML** :
```
image="00" :: a.image    syntax="add" :: d.syntax
```

The same is expressed in LISA by the coding- and syntax-section (cf. chapter 3.2.4).

**LISA** :
```
CODING { 0b00 a }   SYNTAX { "add" d }
```

Both, in nML and LISA, these sections can contain non-terminals referring to further operations. In contrast to nML, in LISA these non-terminals reference by definition the same section of the referenced operation. Concatenation of terminals and non-terminals is expressed by double colons and spaces respectively.

*Specifying structure.* In nML, storages are partitioned into three categories: RAM, register, and transitory storage. Both RAM and register are static storages, i.e. once a value is stored, it is preserved until it is overwritten. In contrast to that, transitory storage keeps its content only for a limited number of machine cycles. The resources are attributed in the model with the information if they are registers, memories, or transitory storages, a type specifier, and (in case they are array structure) the width.

**nML** :
```
let wordsize = 16, type word = int(wordsize),
reg R[8,word]
```

In LISA, the definition of the structure of the processor is based on C/C++ (cf. chapter 3.2.1). All native type specifiers known from C can be used. Besides, user defined data-types are permitted. Moreover, to model arbitrary bit-widths, a special generic integer data-type is provided. The same structure is expressed in LISA as follows:

**LISA** :
```
REGISTER bit[16] R[0..8]
```

In addition to nML, in LISA also the lower array boundary can be specified. However, there is no such thing as transitory storage elements available in LISA.

*Specifying behavior.* Both, in nML and LISA, the execution of operations completely determines the behavior of the machine. Registers and memories represent the state of the machine and an instruction can be seen as a transition function. In nML, the description of behavior is captured in a pre-defined attribute *action*.

**nML** :
```
action = { a=b+c; c=c+1; }
```

It evaluates to a sequence of register-transfers which resemble simple C-statements. Such an allowed statement may be a simple assignment, which can be conditional or unconditional. Also, simple arithmetic operations are permitted. However, this limits the capability of the language to describe complex architectures or the integration of existing IP components in the form of C-code into the model. Describing the behavior on the basis of register-transfers in these cases is counter productive to the idea of simplification by abstraction.

Therefore LISA allows the usage of arbitrary C/C++-code in the behavior-section within operations (cf. chapter 3.2.3). This also includes the usage of function-calls and pointers from within the behavioral code.

**LISA** :
```
BEHAVIOR { a=b+postincr(&c); }
```

*Specifying timing.* When describing multi-cycle operations or pipelined processors, the timing of the target hardware must be described. In nML, the time consumed by functional units is accumulated and annotated at the storages that precede these units, i.e. all computing operations have a duration of zero cycles and only the read from a storage can be delayed by a certain amount of cycles after the write has occurred. The delay value attached to a certain storage describes the time spent at the according stage.

**nML** :
```
reg A[16,word] delay=1
```

However, this limits the nML language to model only architectures with very simple pipelines. Complex pipelining schemes as found in modern architectures, e.g. superscalarity, dynamic dispatching of instructions, etc., cannot be described. Therefore, the LISA language is fundamentally different from nML in this respect. The LISA language provides a powerful generic pipeline model which allows the instantiation of an arbitrary number of pipelines in the model and the assignment of operations to pipeline stage. Besides, the pipeline model provides a set of predefined pipeline control functions giving the user the full control over what is happening in the pipeline (cf. chapter 3.2.5).

The LISA language and its composition of various sections will be presented in more detail in chapter 3. There, special emphasis is laid on the timing model which is key to being able to model state-of-the-art processor architectures.

**Instruction-Set Description Language (ISDL).** The language ISDL of the Massachusetts Institute of Technology (MIT) [59] is primarily targeting the instruction-set description of VLIW processors. In many respects, ISDL is similar to the nML language, however, additionally allows the specification of constraints that determine the valid instructions related to a conflict. Such constraints on instruction-level parallelism (ILP) posed by resource conflicts, etc. are explicitly described in a form of a set of boolean rules, all of which must be satisfied for an instruction to be valid. Pipeline structures cannot be explicitly described in ISDL, which again excludes the generation of cycle-based simulators for pipelined architectures.

From ISDL machine descriptions, software development tools like C-compiler, assembler, linker, and simulator can be generated automatically [38]. Moreover, the possibility of generating synthezisable HDL code is reported [60]. However, as ISDL descriptions mirror the programmer's view of the processor architecture without detailed structural information, it is questionable whether the generated HDL code can be an acceptable replacement of handwritten code. Unfortunately, no results on the efficiency of the generated HDL code are published.

**Valen-C.** Valen-C [61, 39] is an embedded software programming language from Kyushu University. It extends the C programming language by explicit and exact bit-width for integer type declarations. A retargetable compiler, Valen-CC, takes C or Valen-C programs and generates assembly code. Although Valen-C assumes simple RISC architectures, it has retargetability to a certain extent. The target description language driving the compiler retargeting includes the instruction-set description (both behavior and assembly syntax) and the processor data-path width. Besides that, no structural information about the processor is included. The description language used for simulator retargeting is different from the one used to retarget the compiler. Obviously, consistency problems arise due to the use of separate specification languages.

**CSDL.** CSDL is a set of machine description languages used in the Zephyr compiler system developed at the University of Virginia [62]. Zephyr is an infrastructure for compiler writers consisting of several retargetable modules driven by CSDL machine descriptions. CSDL is composed of the following languages: SLED, $\lambda$-RTL, CCL, and PLUNGE. SLED (Specification Language for Encoding and Decoding) [63] describes the assembly and binary representations of instructions and can thus be used to retarget assemblers, linkers, and disassemblers. $\lambda$-RTL describes the behavior of instructions in the form of register transfers [40], while CCL specifies the convention of function calls [64]. PLUNGE is a graphical notation for specifying the pipeline structure. However, it is not clear if cycle-based simulators for pipelined architectures can be generated automatically from PLUNGE. As with Valen-C, the introduction of separate languages describing different views of the architecture (instruction-set, behavior, structure) implies the problem of consistency.

## 1.2 Architecture-Centric Languages

Architecture-centric languages try to derive instruction-set information, valid composition of hardware operations, latencies, and constraints from synthesis oriented machine descriptions. The advantage of this approach is clearly that the same formalism can be used to retarget the software development tools and to synthesize the architecture without changes on the model. However, the most significant drawback is instruction-set simulation speed, as the models contain a huge amount of structural details that is not required even for cycle-based simulation purposes. Consequently, the instruction-set simulators derived are very slow and thus not suited for application software design.

**Mimola.** In Mimola [41], processor architectures are described in terms of a net-list consisting of a set of modules and a detailed interconnect scheme. From that, both the MSSQ [65] compiler and the RECORD compiler system [66] are retargeted. The advantage of Mimola is that it can be used for synthesis, instruction-set simulation, code-generation, and test-generation. However, due to the low abstraction level the performance of the generated simulation tools is poor. This excludes the generated software tools from the usage in the software application design phase.

**COACH.** The COACH [43] system was developed at Kyushu University for the design of ASIPs. It is based on the hardware description language UDL/I [67] in which processors are described at register-transfer (RT) level on a per cycle basis. The instruction-set can be extracted automatically from machine descriptions in UDL/I and used for the retargeting of compiler and simulator. Besides, the UDL/I description is fully synthesizable. However, a major drawback is again the simulation speed of the software simulator.

**AIDL.** AIDL [42] is a hardware description language tailored for the design of instruction-set processors. Unlike hardware description languages like VHDL and Verilog it allows the description of such processor architectures in a very early stage of the design. By this, it enables the designer early in the design process to experiment with different instruction-sets and architectural alternatives to find the best solution for the given application. The level of abstraction introduced by AIDL is called *architecture and implementation level.* At this level, designers only need to consider the instruction-set architecture and the control of instructions without having to consider detailed data-path structures and detailed timing relations.

Timing relations are described in AIDL based on the concept of *interval temporal logic* [68]. For this reason, in contrast to hardware description languages like VHDL and Verilog, AIDL is defined only on discrete time sequences while its behavior is defined on time intervals. Sequentiality and concurrency of execution are expressed by the introduction of *stages* which are basic units in AIDL and usually correspond to the stages in an instruction pipeline. A stage is activated whenever its activating condition is satisfied unless the same stage is under execution.

As with other hardware description languages, the missing formal description of the instructions' binary coding and assembly syntax prohibits the automatic generation of software development tools like code generators or instruction-set simulators. However, it is reported that processor models in AIDL can be simulated in a tailored hardware simulator. Moreover, AIDL descriptions can be translated into synthesizable VHDL code [69].

## 1.3 Instruction-Set and Architecture Oriented Languages

Mixed-level machine description languages describe both the instruction-set and the behavioral aspect of the processor architecture as well as the underlying structure. Some of the languages fitting into this category can retarget software development tools and synthesize the architecture from one sole description. The LISA language underlying this work also fits into this category. However, two significant drawbacks of all other approaches have been overcome by LISA: the ability to abstract on multiple levels of accuracy in the domain of architecture and time (cf. chapter 3.3) and to describe a very broad spectrum of processors with arbitrary architectural characteristics.

**FlexWare.** The FlexWare system from SGS Thomson/Bell Northern Research is a development environment for application-specific instruction-set processors [44] and has been applied in several projects [70]. FlexWare is composed of the code generator CodeSyn [71], assembler, and linker generator as well as the Insulin simulator [72]. Both behavior and structure are captured in the target processor description. The part of the underlying machine description used for retargeting CodeSyn consists of three components: instruction-set, available resources (and their classification), and an interconnect graph representing the data-path structure. The instruction-set description is a list of generic processor macro-instructions to execute each target processor instruction. The simulator uses a VHDL model of a generic parameterizable machine. The parameters include bit-width, number of registers and arithmetic logical units (ALU), etc. The application is translated from the user-defined target instruction-set to the instruction-set of the generic machine. Following that, the code is executed on the generic machine. Again, due to the low abstraction level it is obvious that simulation speed is poor and simulators are not suitable for application design. Besides, the flexibility of the underlying machine description language is limited.

Recently, the FlexWare2 environment [73, 74] has been introduced which is capable of generating assembler, linker, simulator (FlexSim), and debugger (FlexGdb) from the IDL (Instruction Description Language) formalism [75]. The HLL compiler (FlexCC) is derived from a separate description targeting the CoSy [76] framework. Unlike FlexWare, FlexWare2 is not capable of synthesizing the architecture from the underlying machine description.

**MDES.** The Trimaran system [45] uses MDES to retarget compiler backends. MDES captures both structural and behavioral aspects of the target processor architecture and allows the retargeting of HLL compiler, assembler, linker, and simulator. However, retargetability of the cycle-based simulators is restricted to the HPL-PD processor family. It is not clear if synthesis is possible from machine descriptions in MDES.

**PEAS.** Unlike the previous approaches, the PEAS system is not only capable of generating software development tools and synthesizing the architecture, but also of creating an optimal instruction-set for a specific application domain automatically (PEAS-III).

In early work on the PEAS system [46, 77], instruction-set optimization is viewed as the selection of appropriate instructions from a predefined super-set. Here, the given reconfigurable architecture is tuned to each specific application by changing some architectural parameters such as bit-width of hardware functional blocks, register-file size, memory size, etc. The system generates profiling information from a given set of application programs and their expected data. Based on the profiles, the design system customizes an instruction-set from a super-set, decides the hardware architecture (derived from the GCC's abstract machine model [78]), and retargets the software development tools including a HLL compiler [79]. However, selecting instructions from a given super-set cannot always satisfy the demand of diverse applications. In this system, the predefined super-set is fixed, i.e. it cannot be extended by the user. The system is similar to the ASIA system [80] in terms of the inputs and outputs of the design system, however, it differs from ASIA in terms of the machine model and the design method. The PEAS system assumes a sequential (non-pipelined) machine model, whereas ASIA assumes a pipelined machine with a data-stationary control model.

The PEAS-III environment [81, 47] introduces further flexibility to the system. Here, the number of pipeline stages and the behavior of each stage can be varied. Moreover, the instruction formats are flexible and multi-cycle units are available. The system generates both the data-path and the control logic of the processor, a simulation model, and the VHDL description [82]. Resources are chosen from a data-base (FHM-DB), which is parameterized with characteristics like bit-width, algorithm of the operation, etc. The data base provides for each resource a behavioral, an RTL, and a gate-level model. The behavior is based on micro-operation descriptions. The system generates automatically the pipeline control including interlocks. As the system works with a library of predefined components, flexibility in the description of architectures is limited to those components.

**Rockwell Architecture Description Language (RADL).** RADL [48] is an extension of earlier work on LISA [83] that focuses on explicit support of detailed pipeline behavior to enable generation of cycle and phase accurate simulators. However, due to its strong focus on retargetable simulation, information is missing that is required to retarget HLL compilers. Although it seems possible, there are no publications on synthesis from RADL machine descriptions.

**EXPRESSION.** The language EXPRESSION [49, 84] allows for cycle-based processor description based on a mixed behavioral/structural approach. The structure contains a net-list of components, e.g. units, storages, ports, and connections, as well as the valid unit-to-storage or storage-to-unit data transfers. The pipeline architecture is described as the ordering of units which comprise the pipeline stages, plus the timing of multi-cycled units. Operation definitions contain opcode, operands, and formats which are used to indicate the relative ordering of various operation fields. Each instruction is viewed as a list of slots to be filled with operations. Resource conflicts between instructions are not explicitly described, but reservation tables specifying the conflicts are generated automatically and passed to the software tools.

The approach of EXPRESSION aims primarily at the retargeting of software development tools like compiler and simulator [85], however, the generation of synthezisable HDL code seems possible (although to the authors' knowledge nothing has been published). In contrast to the previous approaches, the language supports abstraction from the target architecture on multiple levels of processor model accuracy. To support the processor design process, profiling information can be gathered during simulation run and graphically visualized [86].

**MetaCore.** The MetaCore system [87, 88] is an ASIP development system targeting digital signal processing applications. The system addresses two design stages: design exploration and design generation. In the exploration phase, the system accepts a set of benchmark programs and a structural/behavioral specification of the target processor and estimates the hardware costs and performance of each hardware configuration being explored. Once the hardware design is fixed, the system generates synthesizable HDL code along with software development tools such as C compiler, assembler, linker, and instruction-set simulator. The system does not address the automatic generation of instruction-sets – it rather allows (through profiling) the production of information on the design like performance/cost parameters for a given processor configuration, which guides the designer through the subsequent steps of decisions or choices among alternatives.

The MetaCore system is based on a reconfigurable architecture supporting the general characteristics of various DSP applications, which is parameterized by the MetaCore structural language (MSL) and the MetaCore behavioral language (MBL). The system supports modeling the architecture on multiple abstraction levels. The generated HLL C-compiler is based on the GNU framework [78]. The HDL generator run is divided into two phases. In the macro-block generation, the generator selects macro blocks from a system library and assigns parameters such as size of register/memory and bit-width. In the control-path synthesis, a data-stationary control model [89] is used for pipeline design. Decoders are generated and the opcode is shifted through the pipeline. In the phase of connectivity synthesis, all control inputs and data input/output buses of each functional block are connected to the appropriate control outputs of the decoders and system buses. However, as the approach is based on a library of macro-blocks, retargetability is limited to combinations of blocks found in the library.

**Automatic Synthesis of Instruction-set Architectures (ASIA).** The ASIA design system [80, 90, 91] offers a systematic method to derive the best combination of instruction-sets and micro-architectures from a given set of application benchmarks. It is an extension of the work from Bruce Holmer [92]. The methodology deals with the instruction-set and pipelined micro-architecture co-design, the tradeoff analysis of the design variables performance, code density, instruction word-width, instruction formats, and hardware resources, and the automatic generation of compilation guides to compiler back-ends.

The inputs of the system comprise a set of application benchmarks, an objective function, and a pipelined machine model. The application benchmarks are represented in terms of machine independent micro-operations. The objective function is a user-given function of the cycle-count, the instruction-set size, and the hardware cost. The pipelined machine model specifies the pipeline stage configuration, and the data-path connection. The machine model [93, 94] is based on a pipelined machine with a data-stationary control model [89]. The outputs consist of an instruction-set and the hardware resources that optimize the objective function, as well as compiled codes of the given application benchmarks.

The instruction generation problem is treated by formulating the problem as a modified scheduling problem of $\mu$-operations (MOPs). Each MOP is represented as a node to be scheduled and a simulated annealing scheme is applied for solving the scheduling problem. The ASIA system has been used to specify instruction-sets and micro-architectures in many commercially available processors [95]. However, a strong limitation of the ASIA design system is that code generation is only possible for applications that are considered during architecture design (i.e. in the benchmark suite used as input for the ASIA system).

**Language for Instruction-Set Architectures (LISA).** Previous work on LISA was primarily concerned with the generation of compiled simulators from processor models in the LISA language. In [83, 96], first basic ideas on the language were presented. At that time, the underlying timing model was still based on Gantt charts. As the timing model is the key for being able to model complex architectures, the LISA language was completely revised [97] to the appearance presented in this book (cf. chapter 3.2.5). Besides, the instruction-set model and the way to express operation hierarchy was changed to the pattern of nML.

## 1.4 Other Approaches

Besides the three categories of machine description languages introduced in the previous sections, there are several approaches concerned with particular aspects in the design of programmable architectures which fit in neither of these categories, e.g. BUILDABONG [98, 99], EPIC [100], Partita [101, 102], SAM [103], READ [104], PICO [105], ASPD [106], PRMDL [107], Satsuki [108], ISPS [109, 110], Marion [111], the work of Bajot [112, 106], Engel [113], and Gschwind [114]. Besides, two commercial approaches offering processor design kits deserve mentioning, as they have proven to survive against the established architecture vendors as ARM, Texas Instruments, etc.

**Jazz.** The Jazz system of Improv [115, 116] is composed of a variety of processors. Individual Jazz processors within the meta-core are configurable VLIW processors with two-stage pipeline, four independent 32-bit data-memory buses, and a small configurable amount of instruction memory. For the specified architecture, code generation tools and simulation tools can be generated automatically. Besides, synthesizable HDL code can be derived. However, flexibility is limited to derivatives of the base VLIW architecture.

**Xtensa.** Xtensa [117, 118] of Tensilica Inc. is a configurable RISC-like processor core. The processor consists of a number of base instructions (approximately 70) plus a set of configurable options. These options comprise the width of the register-set, memory, caches, etc. The instruction-set is a super-set of traditional 32-bit RISC instructions. The processor is built around a 5-stage pipeline with 32-bit address space. Xtensa generates synthesizable HDL code and a set of software development tools comprising (GNU based) C-compiler, assembler, linker, simulator, and debugger.

In addition to the base instruction-set, user defined instructions can be added using a proprietary specification language. The extensions to the basic processor core are expressed in the TIE (Tensilica Instruction Extension) language. It expresses the semantics and encoding of instructions. TIE retargets both the software tools as well as the hardware implementation. It allows the designer to specify the mnemonic, the encoding, and the semantics of single cycle instructions. TIE is a low-level formalism similar to the Verilog HDL.

## 2. Motivation of this Work

Obviously, there are many approaches concerned with specific problems in the design process of programmable architectures. This includes the automatic instruction-set design given a set of application programs and constraints, the generation of *efficient* software development tools, and finally the synthesis of the architecture.

However, none of the approaches addresses all aspects associated with the design of a processor on the basis of one sole description of the target architecture. The LISA language and tools were developed to overcome these limitations. The areas of application range from the early architecture exploration, where an abstract model of the architecture is required to perform hardware-software partitioning, to the level of a synthesizable implementation model. Furthermore, a working set of highly efficient software development tools can be generated independently from the level of abstraction. This comprises an assembler, a linker, a fast instruction-set simulator, and a graphical debugger with integrated profiling capabilities. Besides, the software simulator is wrapped in a dedicated interface that allows integration into arbitrary system simulation environments for early system verification.

In short, the main objective of this work was to develop a machine description language and tooling with the following key characteristics:

- ability to abstract on multiple levels of accuracy, starting at the level of the application and going down to the level of the micro-architecture,
- facility to seamlessly refine the processor description from the most abstract model to the implementation model,
- opportunity to generate a complete software development tool-kit on any level of abstraction,
- guaranteeing the generation of production quality software development tools that can be used by the application designer without modification after the processor design is finished,
- capability to generate highly efficient HDL code to enable complete processor generation, and
- enabling easy integration into arbitrary system simulation environments to allow system verification early in the design process.

The following chapters will show how these characteristics are addressed by the LISA language and tools. To make the proof of concept, chapters 5 (*architecture exploration*), 6 (*architecture implementation*), 7 (*software tools for application design*), and 8 (*system integration and verification*) present case studies of real world architectures which the presented methodology was successfully applied to.

# Chapter 3

# PROCESSOR MODELS FOR ASIP DESIGN

The ASIP design methodology presented within the scope of this book is based on machine descriptions in the LISA language. Starting from one sole description of the target architecture the complete processor design can be addressed – architecture exploration, architecture implementation, software tools for application design, and system integration and verification.

The language LISA is aiming at the formalized description of programmable architectures, their peripherals and interfaces. It was developed to close the gap between purely structural oriented languages (VHDL, Verilog) and instruction-set languages for architecture exploration and implementation purposes of a wide range of modern programmable architectures (DSPs and micro-controllers). The language syntax provides a high flexibility to describe the instruction-set of various processors, providing architectural originalities, such as single instruction multiple data (SIMD), multiple instruction multiple data (MIMD), and very long instruction word (VLIW) type architectures [119]. Moreover, processors incorporating heavy pipelining, as seen e.g. in the TMS320C6x of Texas Instruments [120], can be easily modeled. This also includes the ability to describe architectures with complex execution schemes, like e.g. superscalarity [121]. Based on the work of [83] and [97, 122, 123], which was primarily targeting at retargetable simulation, the language was enhanced to support the complete processor design flow [25]. This concerns especially the requirements of the code generation tools (cf. chapter 7.1), the HDL code generator (cf. chapter 6), and the system integration (cf. chapter 8).

This chapter briefly introduces the LISA language by showing its general structuring and composition of different sections. The respective sections contribute to various processor models which are required to address different aspects in processor design. This concerns on the one hand the requirements of the tools for software development, hardware implementation, and system integration. On the other hand, model abstraction is a key requirement to allow seamless refinement of the processor during architecture exploration from abstract specification (application-centric model) to a micro-architectural model (i.e. exact timing and structure).

## 1. LISA Language

The basic structure of the LISA language is adapted from the information found in conventional text-book descriptions of processors, which are frequently called *programmer's manual*. Such programmer's manuals describe:

- the storage elements – the state of the processor, and
- the bit-true behavior of instructions – the transition functions between two states of the processor.

This information is captured in LISA models in processor *resource* declarations and the description of hardware *operations*.

The declared *resources* hold the state of the programmable architecture (captured e.g. in registers, memories, and pipelines) in form of the stored data value and a tag which represents the limited availability of the resources for operation accesses.

*Operations* are the basic objects in LISA. Operation definitions collect the description of different properties of the system, i.e. operation behavior, instruction-set information, and timing. These operation attributes are defined in several sections:

- the *coding* section describes the binary image of the instruction word,
- the *syntax* section reflects the assembly syntax of instructions and their operands,
- the *semantics* section specifies the instruction semantics,
- the *behavior* and *expression* sections describe the transition functions that drive the system into a new state, and
- the *activation* section specifies the timing of other operations relative to the current operation.

During simulation, the execution of these operations drives the system into a new state. The operations can be either atomic or composed of other operations, thus forming an operation hierarchy. Operation hierarchy is built by referencing other operations. In all sections references to other objects are possible. The specification of non-terminal operations on a higher level in the hierarchy is completed by the referenced terminal operations on the lower levels. Figure

3.1 shows a sample operation hierarchy. The operation tree starts with the reserved LISA operation *main* and branches in operation *decode*. Operations *add*, *sub*, *mul*, and *and* terminate the branches as leaves of the tree.

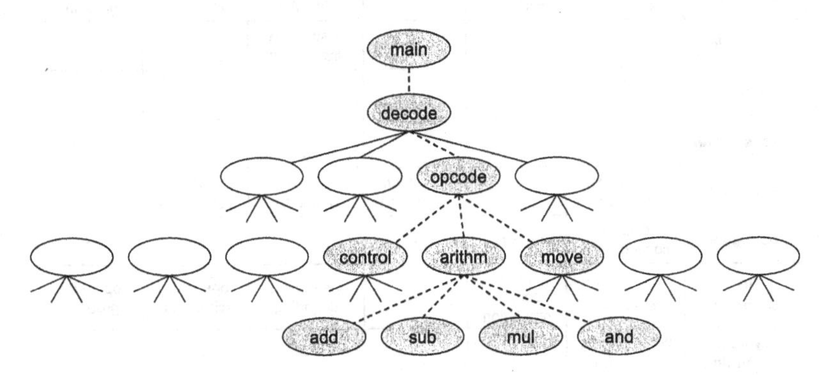


*Figure 3.1.* Operation hierarchy in LISA.

Instructions are formed by composing operations. Generally, the designer is free to determine the abstraction level and modularity of his model based on the operations. For example, one instruction of a processor may be represented by just one operation in case of an instruction-based model or it may be described by a whole set of operations which represent the separate actions between clock cycles in case of a cycle-based pipelined model. A coarse reference of the LISA language can be found in appendix B or requested from LISATek at *info@lisatek.com*.

As already indicated in chapter 2.1.1, LISA incorporates in many respects ideas which are similar to nML. This concerns especially the way the instruction-set is described. As nML does, a LISA model represents an attributed grammar.

## 2. Model Requirements of Tools

The process of generating software development tools, synthesizing, and integrating the architecture requires information on architectural properties and the instruction-set definition as depicted in figure 3.2. These requirements can be grouped into different architectural models – the entirety of these models constitutes the abstract model of the target architecture. The LISA machine description provides information consisting of the following model components.

### 2.1 Memory Model

The *memory model* lists the registers and memories of the system with their respective bit-widths, ranges, and aliasing. The compiler pulls out information on available registers and memory spaces as required during the process of register allocation. Moreover, the memory configuration is provided to perform

| | memory model | resource model | behavioral model | instruction-set model | timing model | micro-architecture model |
|---|---|---|---|---|---|---|
| **HLL-compiler** | register allocation | instruction scheduling | - | instruction selection | instruction scheduling | operation grouping |
| **assembler** | - | - | - | instruction translation | - | - |
| **linker** | memory allocation | - | - | - | - | - |
| **disassembler** | - | - | - | instruction disassembling | - | - |
| **simulator** | simulation of storage | - | operation simulation | instruction decoding | operation scheduling | - |
| **debugger** | display configuration | profiling | - | - | - | - |
| **HDL generator** | basic structure | write conflict resolution | - | instruction decoding | operation scheduling | operation grouping |
| **system integration** | accessability, controlability | - | - | - | - | - |

*Figure 3.2.* Processor model requirements for ASIP design.

object code linking. During simulation, the entirety of storage elements represents the state of the processor which can be displayed in the debugger. The HDL code generator derives the basic architecture structure. For system integration, the pins of the processor and the coupling to external buses are specified. By this, the interface to the model (i.e. accessability of processor resources) is clearly settled.

In LISA, the resource section lists the definitions of all objects which are required to build the memory model. A sample resource section of the ICORE architecture described in [124, 14] is shown in example 3.1. The ICORE architecture also serves as a case study for the architecture implementation step using LISA (cf. chapter 6).

The resource section comprises four types of objects:

- simple resources, such as single registers, buses, flags, and pins as well as vectors hereof such as register files and memories,
- pipeline structures for instructions and data-paths,
- pipeline registers that resemble the data stored in latches between each pipeline stage, and
- memory maps that locate resources in the address space.

The resource section begins with the keyword RESOURCE followed by curly braces enclosing all object definitions. The definitions are made in C-style. As the native C integer data-types are strongly dependent from and limited to the

```
RESOURCE
{
  MEMORY_MAP {
     BUS program_bus: 0x0000 -> 0x1ffff, BYTES(1): ROM[0x0000], BYTES(1);
                      0x2000 -> 0x2ffff, BYTES(1): RAM[0x0000], BYTES(1);
                      0x3000 -> 0xfffff, BYTES(1): /* empty */
  }

  PROGRAM_COUNTER  int             PC;
  REGISTER         signed int      R[0..7];

  DATA_MEMORY      signed int      RAM[0..0x1fff];
  PROGRAM_MEMORY   unsigned int    ROM[0..0x1fff];

  PIPELINE ppu_pipe = { FI; ID; EX; WB };

  PIPELINE_REGISTER IN ppu_pipe
    {
      bit[6]        Opcode;
      short         operandA;
      short         operandB;
    };

  PIN bit[1] interrupt;
  PIN bit[1] endianess;
}
```

*Example 3.1:* Specification of the *memory model* in LISA.

word size of the simulating host, LISA introduces a generic integer data-type *bit*. The scalable *bit* data-type used in LISA supports variable word sizes and provides signed and unsigned functionality. It supports common arithmetic and logical operators as well as data access and modification methods [125].

Resource definitions can be attributed with keywords, like e.g. PROGRAM-_COUNTER, REGISTER, etc. These keywords are not mandatory, but they are used to classify the definitions in order to configure the debugger display. Besides, the register attribute allows the HDL code generator and simulator to recognize clocked resources. The memory map introduced by the keyword MEMORY_MAP maps virtual addresses onto physical resources. The usage of the BUS keyword prior to the address mapping indicates that every access to the respective processor resource (in the example: *ROM*) is automatically redirected onto an external bus (cf. chapter 8).

The resource section in example 3.1 shows the declaration of a simple memory map, program counter, register file, memories, the four-stage instruction pipeline, pipeline-registers, and pins.

## 2.2 Resource Model

The *resource model* describes the available hardware resources and the resource requirements of operations. The instruction scheduling of the compiler is based on this information. The debugger gathers profiling information on resource utilization and operation execution. Furthermore, the HDL code generator is dependent on the resource model for generating the appropriate number of address and data buses to an arrangement of resources as well as for resource conflict resolution.

Besides the definition of all objects, the resource section in a LISA processor description provides information about the availability of hardware resources. By this, the property of several ports, e.g. to a register bank or a memory, is reflected. Moreover, the behavior section within LISA operations (cf. section 2.3) announces the use of processor resources, i.e. the resource requirements of hardware operations. For illustration purposes, a sample LISA code segment taken from the ICORE architecture is shown in example 3.2.

```
RESOURCE
{
  REGISTER    signed int          R[0..7] {
     PORT { READ = 2 OR WRITE = 2 }
  }
  DATA_MEMORY signed int          RAM[0..0x1fff] {
     PORT { READ = 3 XOR WRITE = 3 }
  }
}

OPERATION NEG_RM {
  BEHAVIOR
    USES (READ R[];
          WRITE RAM[];)
    {
      /* C-code */
      RAM[address]  = (-1) * R[index];
    }
}
```

*Example 3.2:* Specification of the *resource model* in LISA.

Resources reproduce properties of hardware structures which can be accessed exclusively by one operation at a time. However, an arrangement of hardware resources, such as registers grouped to a register bank or memories, usually does not allow concurrent access to all elements. Here, dedicated ports for read and write are provided. Introduced by the keyword PORT the declaration is enclosed in curly braces. The number of read and write accesses is specified as a logical conjunction using the operators OR and XOR where the number of reads is assigned to the keyword READ and the number of writes to the keyword WRITE. The example shows the declaration of a register bank and a memory with two and three read and write ports respectively. The definitions differ in the operator used in the declaration. Operator OR indicates that all read and writes stated can be carried out concurrently within one control step while XOR indicates exclusiveness – either reads or writes can be performed.

Furthermore, the section header of the behavior section of the operation *NEG_RM* announces the use of the declared hardware resources for read and write. This is indicated by using the keyword USES in conjunction with the resource name and the information if the used resource is read, written or both (READ, WRITE or READWRITE respectively). As the same information is specified twice – on the one hand explicitly by the USES statement and on the other hand implicitly by the behavioral C-code – the consistency has to be checked manually by the designer. Obviously, the data on used resources could also be extracted automatically from the behavioral C-code by analyzing the code for possible reads and writes to processor resources. This will be addressed in future work.

## 2.3 Behavioral Model

The *behavioral model* abstracts the activities of hardware structures to operations changing the state of the processor for simulation purposes. The abstraction level of this model can range widely between the hardware implementation level and the level of high-level language (HLL) statements.

The BEHAVIOR and EXPRESSION sections within LISA operations describe components of the behavioral model. Here, the behavior section contains pure C/C++-code that is executed during simulation whereas the expression section defines the operands and execution modes used in the context of operations. An excerpt of the ICORE LISA model is shown in example 3.3.

```
OPERATION register
{
  DECLARE { LABEL index; }
  CODING { index=0bx[3] }
  EXPRESSION { R[index] }
}


OPERATION Add {
  DECLARE { GROUP src1,src2,dest = { register }; }
  CODING { 0b010010 src1 src2 dest }
  BEHAVIOR
    {
       /* C-code */
       dest = src1 + src2;
       saturate(&dest);
    }
}
```

*Example 3.3:* Specification of the *behavioral model* in LISA.

Depending on the coding of the *src1*, *src2*, and *dest* field, the behavioral code of operation *Add* works with the respective registers of register bank *R*. As arbitrary C/C++-code is allowed, function calls can be made to libraries which are later linked to the executable software simulator.

## 2.4 Instruction-Set Model

The *instruction-set model* identifies valid combinations of hardware operations and admissible operands. It is expressed by the assembly syntax, instruction word coding, and the specification of legal operands and addressing modes for each instruction. Compiler and assembler can identify instructions based on this model. The same information is used at the reverse process of disassembling and decoding.

In LISA, the instruction-set model is captured within operations. Operation definitions collect the description of different properties of the instruction-set model which are defined in several sections:

- the CODING section describes the binary image of the instruction word,
- the SYNTAX section describes the assembly syntax of instructions, operands, and execution modes, and
- the SEMANTICS section specifies the transition function, i.e. the abstracted behavior of an instruction, in a formalized way.

Example 3.4 shows an excerpt of the ICORE LISA model contributing to the instruction-set model information on the compare immediate instruction (*Compare_Imm*).

```
OPERATION Compare_Imm {
  DECLARE {
    LABEL index;
    GROUP src1, dest = { reg_bank1 || reg_bank2 };
  }
  CODING { 0b10011 index=0bx[5] src1 dest }
  SYNTAX { "CMP" src1 ~"," index ~"," dest }
  SEMANTICS { CMP (dest,src1,index) }
}
```

*Example 3.4:* Specification of the *instruction-set model* in LISA.

The DECLARE section contains local declarations of identifiers and admissible operands. Alternative operations *reg_bank1* and *reg_bank2* are not shown in the figure but comprise the terminal definition of the valid coding and syntax for *src1* and *dest* respectively. The label *index* serves as a local variable for the operation and cross-references information between the sections.

## 2.5 Timing Model

The *timing model* specifies the activation sequence of hardware operations and units. The instruction latency information lets the compiler find an appropriate schedule and provides timing relations between operations for simulation and implementation.

Several parts within a LISA model contribute to the timing model. First, the declaration of pipelines in the resource section. The declaration starts with the keyword PIPELINE, followed by an identifying name and the list of stages.

```
PIPELINE ppu_pipe = { FI, ID, EX, WB };
```
(3.1)

The ordering of stages in the pipeline definition corresponds to their spatial ordering in hardware.

Second, operations are assigned to pipeline stages by using the keyword IN and providing the name of the pipeline and the identifier of the respective stage, such as:

OPERATION *name_of_operation* IN ppu_pipe.EX (3.2)

Third, the ACTIVATION section in the operation description is used to activate other operations in the context of the current instruction. The activated operations are launched as soon as the instruction enters the pipeline stage the activated operation is assigned to. Non-assigned operations are launched in the pipeline stage of their activation. The activation of operations can be seen as the setting of control signals by decoders in hardware.

```
RESOURCE
{
  PIPELINE ppu_pipe = { FI; ID; EX; WB };
}

OPERATION Fetch IN ppu_pipe.FI {
  BEHAVIOR {
    PIPELINE_REGISTER(ppu_pipe, FI/ID).ir = ROM[index];
  }
  ACTIVATION { Decode }
}

OPERATION Decode IN ppu_pipe.ID {
  DECLARE {
     GROUP Instruction = { CORDIC || other };
  }
  CODING { PIPELINE_REGISTER(ppu_pipe, FI/ID).ir == Instruction }
  ACTIVATION { Instruction, Writeback }
}

OPERATION CORDIC IN ppu_pipe.EX  {
  BEHAVIOR {
    PIPELINE_REGISTER(ppu_pipe, EX/WB).ResultE = cordic();
  }
}

OPERATION WriteBack IN ppu_pipe.WB {
  BEHAVIOR {
    R[value] = PIPELINE_REGISTER(ppu_pipe, EX/WB).ResultE;
  }
}
```

*Example 3.5:* Specification of the *timing model* in LISA.

To illustrate this, example 3.5 shows simplified sample LISA code taken from the ICORE architecture. Operations *Fetch* and *Decode* are assigned to stages *FI* and *ID* while operations *CORDIC* and *WriteBack* are assigned to stages *EX* and *WB* of pipeline *ppu_pipe* respectively. For simplicity, operation *other* is not shown.

Here, operation *Fetch* activates operation *Decode* which in turn activates operations *CORDIC* or operation *others* depending on the coding of the instruction. Concurrently, operation *WriteBack* is activated and will be launched in two cycles (in correspondence to the spatial ordering of pipeline stages) in case of an undisturbed flow of the pipeline.

Figure 3.3 shows the pipeline diagram resulting from the LISA code shown in example 3.5 in case of decoding a CORDIC instruction. The arrows denote the activations.

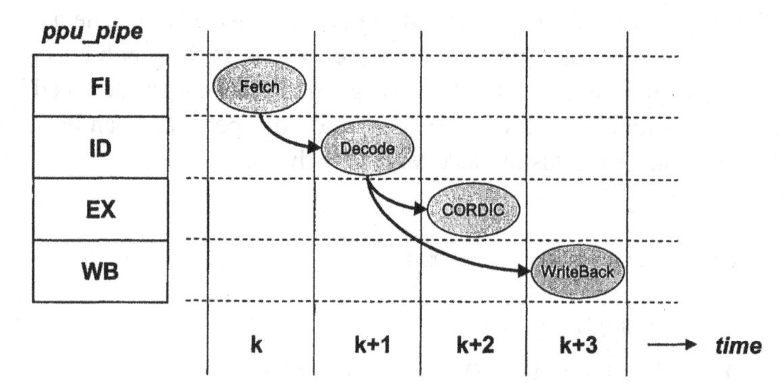


*Figure 3.3.* Execution of an instruction in case of an undisturbed flow of the pipeline.

The actual instruction latency between entering the pipeline and finishing execution not only depends on the number of pipeline stages that have to be passed. Control hazards, data hazards or structural hazards can cause multiple pipeline stalls or even cancellation of the instruction execution. For this reason, in the activation section, pipelines are controlled by means of predefined functions *stall, shift, flush, insert,* and *execute,* which are automatically provided by the LISA environment for each pipeline declared in the resource section. All these pipeline control functions can be applied to single stages as well as whole pipelines, for example:

```
PIPELINE(ppu_pipe,ID/EX).stall();                    (3.3)
```

or

```
PIPELINE(ppu_pipe,ID/EX).flush();                    (3.4)
```

Using this very flexible mechanism, arbitrary pipelines, hazards and mechanisms like forwarding can be modeled in LISA. Figures 3.4a and 3.4b show pipeline diagrams for a *stall* and a *flush* respectively.

In figure 3.4a, it is assumed that in cycle $k+1$ a stall is issued from somewhere in the model for the pipeline register between stages *ID* and *EX*. The stall is taken into account at the next rising clock edge and delays the shifting of the signals resembling activated operations by one cycle. To indicate the stall in the pipeline diagram in figure 3.4a[1], a bubble is inserted. After one penalty cycle, the stall is released and the regular pipeline flow continues.

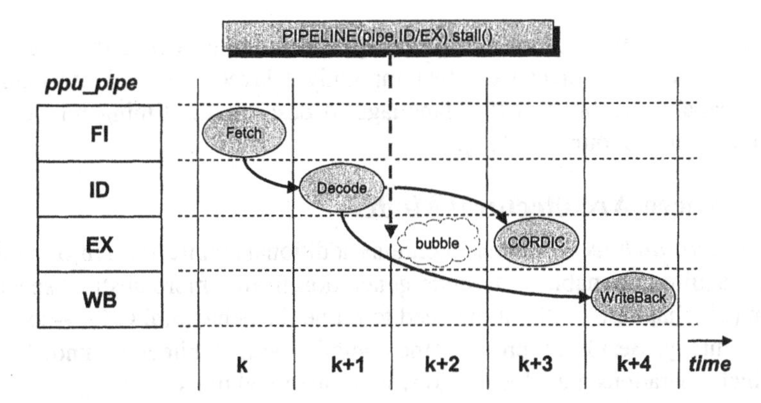


*Figure 3.4a.* Disturbed flow of the pipeline – occurring pipeline stall.

Figure 3.4b shows the pipeline diagram for an occurring flush of the pipeline register between stages *ID* and *EX* in cycle $k+1$. The information of activated operations scheduled for execution in rear stages is deleted and *nop*s (no operations) are inserted instead.

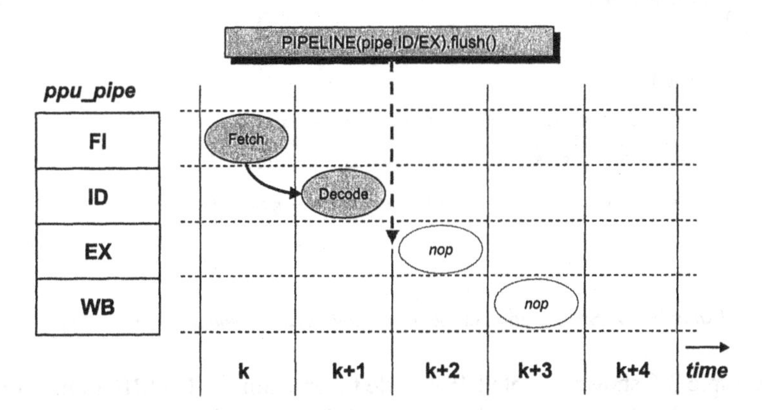


*Figure 3.4b.* Disturbed flow of the pipeline – occurring pipeline flush.

[1]In accordance with the notation used by Hennessy in [126].

The total delay measured in control steps $k_{total}$ between the activation and the execution of a LISA operation is therefore defined by the operation assignment to pipeline stages (spatial delay) and the pipeline control operations stalling the pipeline:

$$k_{total} = k_{spatial} + k_{stall} \tag{3.5}$$

Besides the delay introduced by the spatial ordering of pipeline stages in hardware, the activation can also be temporally delayed. Detailed information on the capabilities of the LISA language to describe the timing relation of operations can be found in [127].

## 2.6 Micro-Architectural Model

The *micro-architectural model* contains additional architectural information that is required to enable HDL code generation from a more abstract specification (cf. chapter 6) without the need to refine the behavioral C/C++-code to RTL accuracy. Besides, it provides the compiler with architectural knowledge on which operations are allocated to which functional units.

For this purpose, hardware operations can be grouped to functional units and data values explicitly set by the decoder outside the behavior sections. This information is later used by the HDL code generator to generate empty frames and the according interconnects to processor resources.

```
RESOURCE {
  UNIT ALU
  {
     Add, Sub, Mul
  }
}

OPERATION Add {
  BEHAVIOR
  SET (PIPELINE_REGISTER(pipe, ID/EX).operand_1 = reg16,
       PIPELINE_REGISTER(pipe, ID/EX).operand_2 = R[2],
       PIPELINE_REGISTER(pipe, ID/EX).operand_3 = 0b101;)
  {
    // Behavioral C-code (is discarded for HDL code generation !)
  }
}
```

*Example 3.6:* Specification of the *micro-architectural model* in LISA.

Example 3.6 shows sample LISA code taken from the ICORE architecture. Apart from the definition of all objects and their availability (cf. *memory* and *resource model*, sections 2.1 and 2.2) the resource section also contains the definition of functional units. Operation grouping to functional units is formalized using the keyword UNIT followed by an identifying name and a list of operations assigned to that unit. The example shows the definition of the unit *ALU* containing hardware operations *Add*, *Sub*, and *Mul*.

As the behavioral C/C++-code is discarded during HDL code generation, only the information specified in the SET statement is taken into account. The explicit setting of data values is formalized in the header of the behavior section. Here, simple assignments of immediate values and contents of processor resources to pipeline registers are permitted.

# 3. Abstraction Levels

Besides the information provided by the LISA language on different processor models, one of the key aspects in architecture development is the ability to abstract on multiple levels of accuracy. However, it is of course mandatory that a working set of software development tools can be successfully generated independently of the abstraction level. Especially in the architecture exploration phase, seamless refinement of the model is important to shorten design cycles while keeping the models and tools consistent.

Abstraction can take place in two different domains: *architecture* and *time*. Both domains are closely correlated since it is unlikely that spatial accuracy is set to the level of micro-architecture implementation whereas temporal accuracy is that of the boundary of HLL statements.

## 3.1 Abstraction of Architecture

Abstraction of the architecture concerns the granularity in which the hardware structure and behavior is described. Figure 3.5 shows a coarse classification of different abstraction levels in the architecture domain.

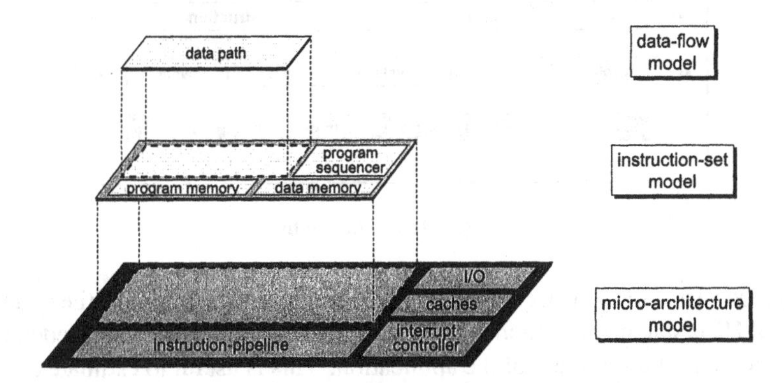


*Figure 3.5.* Abstraction of architecture.

At a very early stage of the design, architecture development starts with an un-timed functional specification of the algorithm that is to be mapped onto the processor architecture. At this point the designer realizes a bit-true version of the processor's data-path abstracting the granularity of hardware resources and behavior to the level of a pure data-flow model of the target architecture.

When the data-flow model meets the requirements the model is refined and the instruction-set is introduced. For this purpose, hardware granularity and behavior are enriched by adding data and program memories to the model as well as an instruction sequencer. This model (the instruction-set model) can then be used to explore the instruction-set by profiling the application on the target architecture. Once the instruction-set is fixed, different micro-architectural implementations are examined that implement the instruction-set. For this, structure and behavior are once again refined by adding hardware details, such as pipelines, peripherals, caches, I/Os, and interrupt controllers. At this level representing a micro-architectural model of the target architecture, the model contains enough structural and behavioral details to co-simulate or even synthesize hardware structures. Chapter 5 will illustrate a stepwise refinement of the LISA model of the ICORE architecture.

## 3.2 Abstraction of Time

As it is necessary to vary the granularity in which structure and behavior are described, the same applies to abstraction of time. Figure 3.6 shows different levels of abstraction of time.

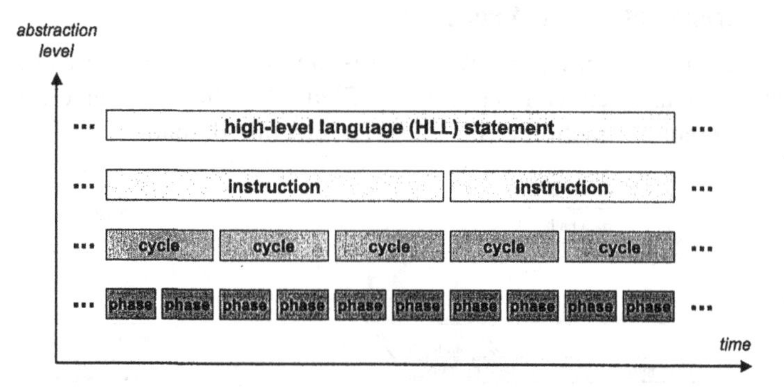


*Figure 3.6.* Abstraction of time.

Temporal accuracy at an early stage of the design can be limited to the boundaries of HLL statements which only represents an estimate for time needed for the execution of a fragment of the application. This is useful to estimate e.g. if real-time constraints can be fulfilled and is frequently referred to as cycle-approximate models. Time can then successively be refined to the boundaries of processor instructions (instruction-based model). For implementation and co-simulation purposes, time can further be refined to the level of cycles or even phases within cycles (cycle/phase-based models).

## 3.3 Limitations

Common to all processor models realized in LISA is the underlying zero-delay model. This means that all transitions are provided correctly at each control step. Control steps may be clock phases, clock cycles, instruction cycles, or even higher levels (cf. section 3.2). Events between these control steps are not regarded.

One of the strengths of the LISA language is the underlying timing model, which allows modeling of arbitrary complex pipelines and execution schemes. As explained in section 2.5, it relies on the specification of pipeline stages, operation activation, and the assignment of operations to pipeline stages. However, this also introduces a few drawbacks: on the one hand, the implicit specification of latency information, which is of key importance to enable cycle-accurate simulation and HDL code generation, makes the extraction of reservation tables required for compiler retargeting difficult. On the other hand, the static assignment of operations to pipeline stages prohibits to model an architectural characteristic which is sometimes found in CISC architectures: out-of-order execution of instructions. Here, instructions in the pipeline can overtake each other. Again, a possible fallback solution is to model these cases in pure C/C++. In this case, the generic pipeline model could still be used, whereas the pipeline control operations normally used to shift values concerning activated operations through the pipeline would have to be implemented manually. By this, the concurrent shifting of all activated operations through the pipeline could be overridden.

In addition to these restrictions, there are few possible enhancements to ease processor model refinement. This especially concerns the transition from instruction-based to cycle-based models (cf. chapter 9).

# 4. Concluding Remarks

In this chapter the language for instruction-set architectures (LISA) was introduced. It was illustrated that two major requirements are fulfilled by the language to address the complete design flow of a programmable architecture: the ability to retarget and to abstract.

The requirements introduced by the tools for retargeting are addressed in LISA in different processor models of the target architecture – *resource*, *memory*, *behavioral*, *instruction-set*, *timing*, and *micro-architectural model*. Model abstraction can take place both in the domain of architecture and time on multiple levels.

# Chapter 4

# LISA PROCESSOR DESIGN PLATFORM

The LISA processor design platform (LPDP) is an environment supporting the complete design flow of ASIPs. This comprises the automatic generation of software development tools for architecture exploration, hardware implementation, software development tools for application design, and co-simulation interfaces from one sole specification of the target architecture in the LISA language. Figure 4.1 shows the areas of application of the design environment.

The first quadrant addresses the *architecture exploration phase*. Within this phase, the target architecture is tailored to the application domain with regard to certain design constraints, e.g. throughput and resource utilization. Beginning at an abstract level, the architecture can be refined to the level of the micro-architectural implementation (cf. chapter 3.3). However, independently of the model accuracy, a working set of software development tools can be generated to benchmark the architecture.

The second quadrant aims at the *architecture implementation phase*. Here, the resulting LISA model from the exploration phase is considered as a basis for the generation of synthesizable HDL code. It is obvious that the required abstraction level of the LISA model capable of synthesis is that of micro-architectural and cycle accuracy.

The third quadrant addresses the *application software design phase*. For this purpose, a set of production quality software development tools is provided to enable the software designer to comfortably program the architecture. While in the architecture exploration phase the designer demands primarily flexibility of the tools and visibility of the operation of the architecture, in the application design phase speed and functionality of the tools is most important.

To enable the verification of the architecture in conjunction with the rest of the system, the fourth quadrant of the LISA processor design platform aims at the *system integration and verification phase*. To cope with the requirements of common system simulation environments, such as [128, 129, 130], dedicated interfaces are provided by the environment to allow seamless integration.

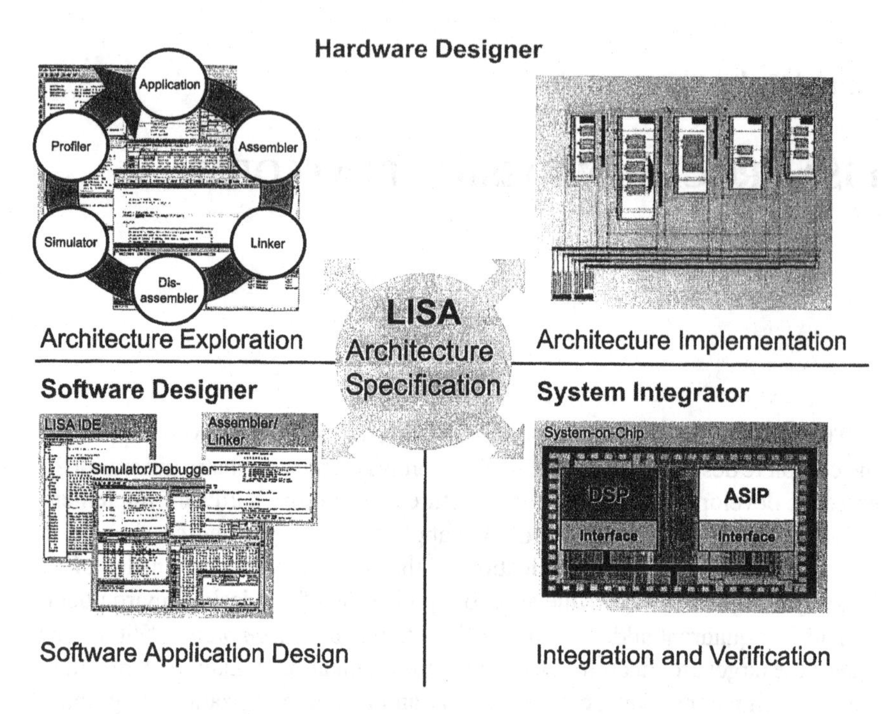


*Figure 4.1.* LISA processor design platform.

This chapter introduces the different areas of application addressed by the LISA processor design platform. It is organized by the different groups of people associated with the processor design process: hardware designer, software designer, and system integrator (cf. chapter 2). The first and second quadrant (architecture exploration and architecture implementation, respectively) are assigned to the hardware designer, while the software application design is assigned to the software designer. The system integration and verification is linked with the group of system integrators.

## 1. Hardware Designer Platform

In the architecture exploration and implementation phase, the hardware designer has to deal with three different problems:

- the design of the target architecture (i.e. realization of the architectural specification),
- the building of software development tools, and

- the realization of a synthesizable model in a hardware description language like VHDL or Verilog.

During the phase of architecture exploration, the architecture is optimized for the target application domain. This optimization process is carried out by benchmarking the application on different architectural alternatives. Obviously, benchmarking can only be efficiently carried out at the presence of a complete software development tool-suite. For this reason, software development tools consisting of at least an assembler, a linker, and a software simulator have to be available to the designer from the very beginning of the exploration phase.

Starting with a first draft of the architecture ("educated guess"), the application is assembled, linked, and then simulated on the software simulator to get quantitative data on the performance of the architecture and application. At this stage of the design, the evaluation results will provide information about the suitability of the chosen instruction-set, the resource utilization, and the execution speed. If the posed design goals are not fulfilled, the architecture is changed according to the benchmarking results and a new iteration of benchmarking the application on the target architecture is initiated. This process is repeated, until the design goals are met and a best-fit between application and architecture is obtained (see figure 4.2).

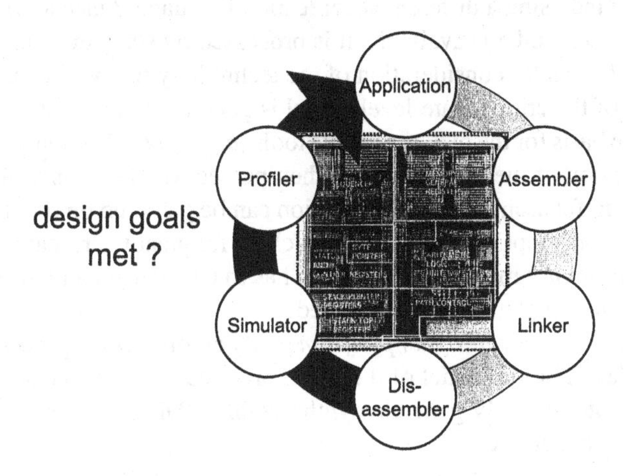


*Figure 4.2.* Architecture exploration by iterative refinement of the architecture.

It is obvious that every time the architecture is changed all tools need to be adapted as well. However, changing the tools manually is a very tedious, lengthy, and error-prone process as the tools are usually written by hand in a high-level language like C. Besides, consistency among the tools and between the tools and the underlying architecture specification is critical. A major problem here is verification.

The *hardware designer platform* of the LISA processor design platform automates the realization of the software development tools. Jointly with the architectural specification, a LISA model is realized which parameterizes the software development tools provided by the LISA environment. The suite of tools consists of the following components:

- retargetable assembler which translates text-based instructions into object code for the target architecture,
- retargetable linker which generates an executable application from multiple object files,
- retargetable disassembler which reads the executable application and extracts the assembly syntax for display in the debugger, and
- retargetable instruction-set architecture (ISA) simulator which executes the application and provides extensive architecture and application profiling capabilities, such as instruction execution statistics and resource utilization.

Following the architecture exploration phase, the architecture needs to be implemented in a hardware description language like VHDL or Verilog. At this point there is a break in the design flow as the processor model has to be re-implemented using a different specification language. Once a synthesizable implementation model is available, it is processed by standard synthesis tools, e.g. [30, 131]. Under consideration of the technology that will be used for the production of the chip, a gate-level model is generated. This gate-level model serves as the basis for the place and route tools (e.g. [132, 133]) that help to build the final version of the chip, which is then passed to a fab. Only on the gate-level, meaningful architectural information can be gathered on required silicon area, power consumption, and maximum clock frequency (critical path). If the design goals are not met, the architecture has to be changed and the process of exploration and implementation repeated.

The LISA *hardware designer platform* partly automates the process of building an implementation model of the target architecture. This concerns the capability to automatically generate synthesizable HDL code for the following parts of the architecture:

- control-path,
- instruction decoder,
- pipeline and pipeline controller,
- processor resources, such as registers and signals, and
- frames for functional units with their interconnects to processor resources.

For the data-path, hand-optimized code has to be inserted manually into the generated frames of the HDL model. This approach has been chosen as the data-path typically represents the critical part of the architecture in terms of power consumption and performance that needs to be optimized by hand and is often realized in full custom design [134].

The LISA-based architecture exploration and implementation flow is shown in figure 4.3. The LISA description is processed by the LISA processor compiler which generates C++ source-code for the software development tools and VHDL code for the implementation model. The generated C++-code is then compiled with a native C++-compiler on the host platform (not shown in figure 4.3).

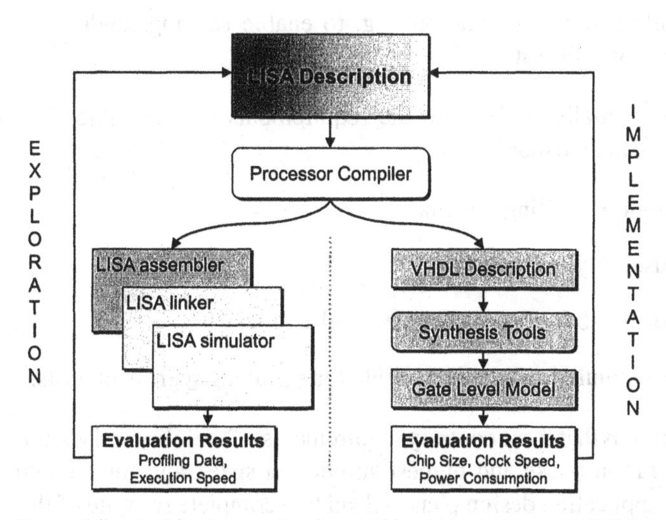


*Figure 4.3.* Architecture exploration and implementation using LISA.

It is obvious that deriving both – the software tools and the hardware implementation model – from one sole specification of the target architecture in the LISA language has significant advantages: only one model needs to be maintained, changes on the architecture are applied automatically to the software tools and the implementation model and the consistency problem between the architectural specification, the software tools, and implementation model is reduced significantly. As the data-path has to be specified twice – on the one hand in the LISA model and on the other hand in the implementation model – there is still a need to verify those portions of the models against each other.

## 2. Software Designer Platform

When the architecture is fully specified by the hardware designers, a set of production quality software development tools is needed to enable the application designers to comfortably program the architecture. As in the traditional processor design flow the exploration tools are realized manually by the hardware designer, the tools need to be re-written for the application design phase. The reason for this is founded in the different areas of application of the tools. The architecture designer requires:

- full visibility of the architecture,
- architecture and application profiling information,
- cycle-accuracy in the domain of time and micro-architectural accuracy in the domain of architecture, and
- flexibility during simulation, e.g. to enable running randomly generated instructions for testing.

This is partially opposed to the requirements of the application software designer, who demands

- application profiling information,
- easy usability of the tools,
- graphical debugging capabilities, and
- highest simulation speed to enable long running regression tests.

It is obvious that these contrary requirements – flexibility and visibility during the exploration phase and highest simulation speed and functionality during software application design phase – lead to a complete re-write of the software development tools. Re-implementation of the software tools for application design takes place on the basis of architectural information which is passed between the hardware designers and the application software tool designers in the form of a textual specification. However, using a textual specification induces the problem of consistency. It is a well known fact that it takes several revisions of the software development tools for application design until the tools really match the hardware implementation.

The *software designer platform* of the LISA processor design platform automates the realization of software development tools. To cope with the requirements in terms of functionality and speed in the software application design phase, the tools are an enhanced version of the tools generated during the architecture exploration phase (cf. section 1). Moreover, a graphical debugger visualizing the simulation progress is provided as well. Besides the generated software development tools, the software designer platform provides an integrated development environment (IDE) which allows comfortable maintenance

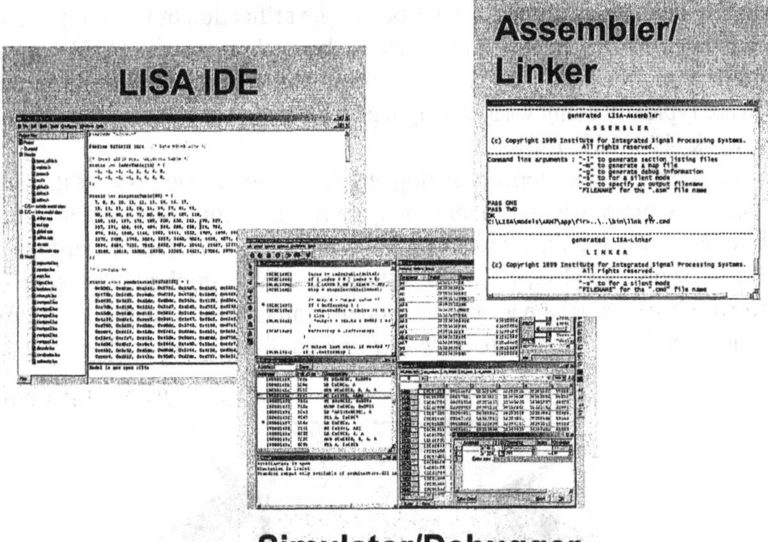


*Figure 4.4.* Application development tools generated from LISA.

of projects. Figure 4.4 shows the generated tools for the software application design.

The enhancement of the tools concerns primarily the generated simulators. They are enhanced in speed by applying the compiled simulation principle – where applicable – and are faster by one to two orders in magnitude than the tools currently provided by architecture vendors (cf. chapter 7.4). As the compiled simulation principle requires the program memory not to be modified during the simulation run, this holds true for most DSPs. However, for architectures running the program from external memory or working with operating systems which load/unload applications to/from internal program memory, this simulation technique is not suitable. For this purpose, the more flexible, though slower, interpretive simulation technique is provided as well.

## 3. System Integrator Platform

Once the processor software simulator is available, it must be integrated and verified in the context of the whole system. Today, typical single chip electronic system implementations include a mixture of micro-controllers, digital signal processors as well as shared memory, dedicated logic and on-chip communication. Driven by the ever increasing hardware and software design complexity, components from various design teams and third parties (intellectual property blocks) are employed. Due to the heterogeneity of these components and the drastically increased number of gates per chip, verification of the complete system has become the critical bottleneck in the design process [135]. System integration environments integrate hardware and software design techniques, which are typically using various languages, formalisms, and tools into a single framework [128, 129, 130].

Figure 4.5 shows a system consisting of heterogeneous components that are to be integrated into one environment for system simulation. Here, heterogeneity not only concerns the different formalisms and tools – VHDL or Verilog with simulation tools like [136, 137, 138, 139] for the HDL implementation model, software debuggers for the processor models, and C/C++ debuggers for purely functional models – but also models working on different abstraction levels.

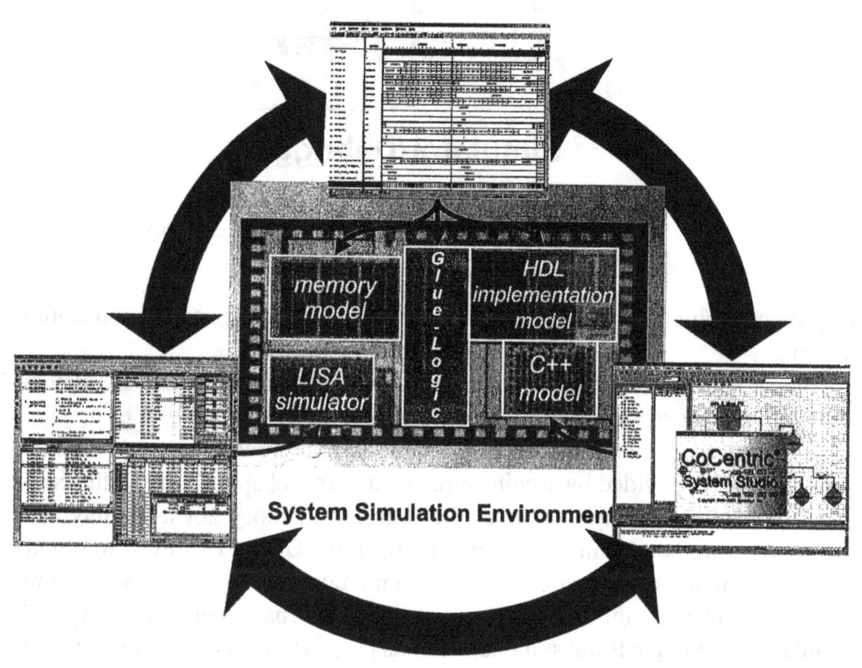


*Figure 4.5.* System simulation environment consisting of heterogeneous components.

In order to support the system integration and verification, the *system integrator platform* of the LISA processor design platform provides a well-defined application programming interface (API) to connect the instruction-set simulators generated from the LISA specification with other simulators. The API allows to control the simulator by stepping, running, and setting breakpoints in the application code and by providing access to the processor resources. The LISA environment does not provide a system simulation environment – it rather enables seamless integration of its tools into existing environments. Chapter 8 will show the system integration into the commercially available CoCentric System Studio environment of Synopsys [26].

## 4. Concluding Remarks

In this chapter the LISA processor design platform was introduced. It addresses the complete processor design flow, starting from abstract specification during the architecture exploration phase and going as far as processor implementation, application software tool generation, and seamless system integration.

It was shown, that basing all phases of the processor design on one sole specification of the target architecture in the LISA language accelerates the processor design process enormously and reduces the consistency problem between the phases significantly. The following chapters will illuminate the four quadrants of the LISA processor design platform more closely – architecture exploration, architecture implementation, tools for application software development, and system integration and verification.

Chapter 5

# ARCHITECTURE EXPLORATION

Application-specific programmable architectures are considered the most promising design paradigm to cope with the ever increasing complexity of system-on-chip (SOC) designs [140]. The appealing idea is to define a programmable platform tailored to a specific application domain at the price of restricted general-purpose properties. Such a platform is able to incorporate diametrical requirements like high processing power, flexibility and reusability together with moderate power consumption and hardware costs.

However, this puts a heavy burden on the ASIP designer to compose a capable platform from a huge design space for a given application domain. The goal of the LISA-based processor design methodology is to guide the designer through the whole processor design flow – from the algorithmic specification of the application down to the implementation of the micro-architecture. At every stage of this flow the designer maintains an abstract model of the evolving target architecture written in the LISA language. From this LISA model, a working tool-set supporting the evaluation tasks at the current stage of the design flow can be generated automatically.

This chapter introduces the first quadrant of the LISA processor design platform, which is concerned with the architecture exploration (cf. chapter 4.1). The process of model refinement is carried out exemplarily for a simple CORDIC application to be mapped onto an application-specific processor architecture. Architectural and application profiling capabilities of the generated tools are shown, which are required to optimize and evaluate the architecture during the different design phases. However, it is not the primary intension of this chapter to develop a *production ready* processor architecture or to present formal cost models to evaluate the architecture for its suitability for the target application. It is rather the goal to introduce different processor models and the respective tools that support the designer in making certain design decisions on varying levels of abstraction.

## 1. From Specification to Implementation

The proposed processor design methodology sets in after the algorithms, which are intended for execution on the programmable platform, are selected. The algorithm design is beyond the scope of LISA and is typically performed in an application-specific system level design environment, like e.g. COSSAP or CCSS [141, 26] for wireless communications or OPNET [142] for networking.

The architecture-specific processor design flow can be classified into six phases, which are characterized by their underlying processor model:

- high-level language (HLL) algorithmic kernel model,
- parallelized HLL algorithmic kernel model,
- data-path model,
- instruction-based model,
- cycle-based model, and
- register transfer level (RTL) micro-architectural model.

Figure 5.1 visualizes the different steps of architecture exploration. The arrow down indicates increasing architectural refinement while the loops represent iterations required to optimize the architecture on the respective level of abstraction.

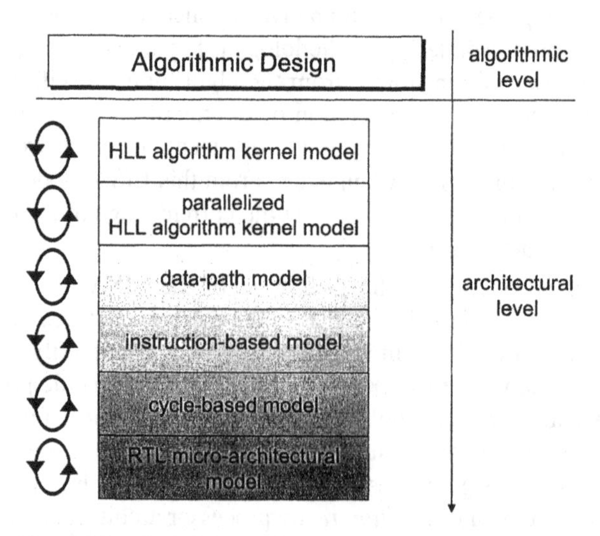


*Figure 5.1.* Classification of steps in the design flow of ASIPs.

The outcome of the algorithmic exploration is a purely functional specification usually represented by means of an executable prototype written in a high-level language like C, together with a requirements document specifying cost and performance parameters.

**Phase 1 – HLL algorithmic kernel model.** Starting point on the application side is the algorithmic kernel of the application, i.e. those parts of the application which are candidates for application-specific hardware support in the ASIP. On this level, execution statistics of the application code are gathered, where the cumulative percentage for the execution of each HLL statement becomes visible. The HLL execution statistic is of course only of limited precision with respect to the final processor, but is valuable to give an initial idea of the required resources and special-purpose processor extensions which have to be employed into the processor to meet the specified performance requirements.

**Phase 2 – Parallelized HLL algorithmic kernel model.** In the next step, the appropriate processor category is selected. Here, trade-offs between parallel versus sequential execution schemes are examined and the required degree of instruction-level parallelism (ILP) is defined. To get sensitive estimations on the performance gain attainable with ILP at this early stage of the design flow, the application code can be modified enabling the parallel execution of HLL statements. Now, the profiling results take parallel resources into account and by trying out different degrees of parallelism, the designer is directed towards the optimal architecture for the given application.

**Phase 3 – Data-path model.** At this point of the application refinement, performance critical parts of the application and the required degree of ILP are identified. Based on that, the data-path of the programmable architecture on the assembly instruction-level can be defined. Starting from an arbitrary basic instruction-set, the programmable architecture can be enhanced with parallel resources, special-purpose instructions, and registers in order to improve the performance of the considered application. At the same time, the algorithmic kernel of the application code is translated into assembly by making use of the specified ILP and the special-purpose instructions. By iteratively profiling and modifying the programmable architecture in cadence with the application, both converge against the specified performance requirements.

**Phase 4 – Instruction-based model.** After the processing intensive algorithmic kernels are considered and optimized, the architecture is completed by adding instructions which are dedicated to the low speed control and configuration parts of the application. These parts usually represent major portions of the application in terms of code amount but have only negligible influence on the overall performance.

**Phase 5 – Cycle-based model.** Until this phase, optimization was only performed with respect to the software related aspects, while neglecting the influence of the micro-architecture. Therefore, architectural information about pipelines and the instruction distribution over several pipeline stages has to be taken into account. If the architecture does not provide automatic interlocking mechanisms, the application code has to be revised to take pipeline effects into account. Here, it has to be verified that the cycle-based processor model still satisfies the performance requirements.

**Phase 6 – RTL micro-architectural model.** In a final step, the architectural behavior has to be refined to the level of register transfers. After implementing the dedicated execution units of the data-path, meaningful numbers on hardware costs and performance parameters (e.g. design size, power consumption, clock frequency) can be derived by running the HDL processor model through the standard synthesis flow. On this high level of detail the computational efficiency of the architecture can be optimized by applying different implementations of the data-path execution units.

Of course, processor design is never performed in the idealized top-down fashion as outlined above. The designer rather has to iteratively cycle through the phases to tailor the architecture to the application. The frequency and the extent of necessary iterations mainly depends on the skills and experience of the designer, since decisions at early stages in the flow must take the following refinement steps into account.

## 2. Architecture Exploration Using LISA

The proposed ASIP design flow is only useful in conjunction with a methodology supporting this flow. The LISA language and tooling can be used from the very beginning of the processor design to profile the application and the processor model can then be seamlessly refined to the implementation level. As explained in chapter 3.3, LISA allows modeling the architecture on various levels of abstraction. Independently on the level of abstraction, a working set of tools can be generated automatically. Figure 5.2 shows the LISA-based processor design flow and the exploration results gathered on the respective levels using the LISA tool-suite. Besides the processor model refinement, also the application refinement during the different processor design phases is displayed.

This chapter will go step by step through the six phases of the design of application-specific processors and present sample LISA models supporting the respective design stage. The target application to be mapped onto an ASIP is a CORDIC angle calculation, which is a simplified fragment of the version of the real application implemented in the ICORE architecture described in [124, 14]. The ICORE architecture also serves as a case study for the architecture implementation step using LISA (cf. chapter 6).

| | application | LISA model | exploration result |
|---|---|---|---|
| 1 | HLL algorithm kernel | generic HLL model | HLL operation execution statistics |
| 2 | parallelized HLL algorithm kernel | abstract resource model | ILP performance gain |
| 3 | assembly algorithm kernel | data-path model | ISA accurate profiling (data) |
| 4 | assembly program | instruction-based model | ISA accurate profiling (data+control) |
| 5 | revised assembly program | cycle-based model | cycle accurate profiling (data+control) |
| 6 | assembly program | RTL micro-architectural model | HW cost + timing |

application and model refinement

*Figure 5.2.* LISA-based ASIP development flow.

## 2.1 Description of the CORDIC Algorithm

The CORDIC (*CO*rdinate *R*otation *DI*gital *C*omputer) trigonometric computing technique [143] can be used to solve either set of the following equations:

$$x = k(x_0 \cos\alpha - y_0 \sin\alpha) \tag{5.1}$$

$$y = k(x_0 \sin\alpha + y_0 \cos\alpha) \tag{5.2}$$

or

$$r = k\sqrt{x^2 + y^2} \tag{5.3}$$

$$\vartheta = \tan^{-1}\frac{y}{x}, \tag{5.4}$$

where $k$ is an invariable constant. Using the basic trigonometric relation

$$\cos^2\alpha = \frac{1}{1 + \tan^2\alpha} \tag{5.5}$$

yields to

$$x = \frac{1}{\sqrt{1 + \tan^2\alpha}}(x_0 - y_0 \tan\alpha) \tag{5.6}$$

$$y = \frac{1}{\sqrt{1 + \tan^2\alpha}}(x_0 \tan\alpha + y_0)\,. \tag{5.7}$$

The term $\tan^2 \alpha$ determines the influence of the angle to the new coordinates. There are two computing modes, for which the CORDIC algorithm can be used – rotation and vectoring. In the *rotation* mode the coordinate components of a vector and a rotation angle are given. Using the rotation mode, the components of the original vector, after the rotation through the given angle, are computed. In *vectoring* mode, the coordinates of a vector are given and the angular argument and the magnitude of the original vector are determined. The whole algorithm is based upon the following iterations, which do not apply to a full rotation, but perform a rotation in several steps:

$$x_{n+1} = x_n - d_n y_n 2^{-n} \tag{5.8}$$

$$y_{n+1} = y_n + d_n x_n 2^{-n} \tag{5.9}$$

$$z_{n+1} = z_n - d_n \arctan 2^{-n} \,. \tag{5.10}$$

The terms $\arctan 2^{-n}$ are pre-computed and stored in a lookup table within a small part of the memory. Small angle values which are sufficient for the approximation $\tan \frac{x}{2} \approx \frac{x}{2}$ are not stored in this table. In the rotation mode of the CORDIC algorithm, $d_n$ is equal to the sign of $z_n$ (namely $+1$ when $z_n \geq 0$, else $-1$). This leads to the following results:

$$x_{n+1} = x_n \mp y_n 2^{-n} \tag{5.11}$$

$$y_{n+1} = y_n \pm x_n 2^{-n} \tag{5.12}$$

$$z_{n+1} = z_n \mp \arctan 2^{-n} \,. \tag{5.13}$$

These equations require only addition, subtraction and shift operations in hardware, because the arctan values are being looked up in the pre-computed arctan-table. No multiplication, which is hard to implement efficiently, is needed.

```
/********************************/
/*    Computing sin and cos     */
/********************************/

/* Initialization  of variables */
/* C => precalculated constant  */
/* alpha => angle               */

X = C
Y = 0
Z = alpha
N = 10

/* Begin CORDIC calculation */

for i = 0 to N
    dx = X / 2^i
    dy = Y / 2^i
    dz = atan (1 / 2^i)
    if Z >= 0 then
        X = X - dy
        Y = Y + dx
        Z = Z - dz
    else
        X = X + dy
        Y = Y - dx
        Z = Z + dz
    endif
next
```

*Example 5.1:* Pseudo C-code of the CORDIC algorithm.

Example 5.1 shows pseudo C-code for the CORDIC calculation. This code is the starting point of the architecture design presented in the following sections. As already indicated at the beginning of this chapter, it is neither the goal to develop a highly optimized architecture nor to introduce formal cost models on the respective levels of abstractions to make design decisions of the architecture. The refinement steps are meant to indicate, how the capabilities of the LISA language to model the target architecture on different levels of abstraction in cadence with the respective tooling can be used to support such cost models. Especially the diverse profiling capabilities of the generated processor debugger are of great value for this purpose.

However, in order to carry out a sensible architecture exploration, a design goal has to be specified. In the following sample processor model refinement, the goal is to minimize cycle count, i.e. to maximize throughput and thus performance, while minimizing the size of the application program. Performance is quantified by counting the number of control steps required to execute the application program while program size is quantified by counting the number of bytes required to store the application in memory. Obviously, this size can only be determined once the instruction-set and thus the binary coding of instructions has been introduced to the processor model.

## 2.2 Generic HLL Model

The first phase within the design process of an ASIP is primarily concerned with the examination of the application to be mapped onto the processor architecture. Here, critical portions of the application need to be identified that will later require parallelization and specific hardware acceleration.

On this level, a generic model in LISA is realized, whose instruction-set implements an ANSI C subset. Example 5.2 shows sample LISA code of a generic architecture model while example 5.3 shows the application to be executed on the model. The variables used within the application code are now incorporated in the LISA model as resources. Besides, a program counter is instantiated to enable control-flow. The initialization of the loop counter $i$ and the number of iterations $N$ is performed in the reserved operation *reset*. This operation is executed automatically by the simulator when the processor model is instantiated. Furthermore, the values for the arctan are precalculated and stored in an array. As these values are independent from the angle to be calculated, they will later be put as constant values in a lookup table that can be stored in the architecture's ROM. Moreover, LISA operations are realized implementing the respective operations used within the application code – addition, subtraction, shift, and simple control-flow. For simplicity, the operations mirroring the source and destination variables of the application code (*dest*, *src1* and *src2*, respectively) are omitted in the example.

```
/************************************/
/*  LISA model implementing virtual */
/*  machine executing ANSI-C subset */
/************************************/

/* Declaration of variables */
RESOURCE {
  int X, Y, Z;
  int dx, dy, dz;
  int arctan[0..7];
  char i, N, cnt;
  PROGRAM_COUNTER PC;
}

/* Initialization of variables */
OPERATION reset {
  BEHAVIOR {
     /* Init dx,dy,dz,N and arctan */
     N = 10; dx = dy = dz = 0;
     arctan[0] = ... ; arctan[10] = ...;
  }
}

/* Main operation */
OPERATION main {
  DECLARE {
    GROUP operations =
         { ADD || SUB || ... } }
  SYNTAX { operations }
  BEHAVIOR {
    operations(); PC++;
  }
}

/* Operation + */
OPERATION ADD {
   DECLARE {
      GROUP dest, src1, src2 = { ... }; }
   SYNTAX { dest "=" src1 "+" src2 ";" }
   BEHAVIOR { dest = src1 + src2; }
}

/* Operation - */
OPERATION SUB {
   DECLARE {
      GROUP dest, src1, src2 = { ... }; }
   SYNTAX { dest "=" src1 "-" src2 ";" }
   BEHAVIOR { dest = src1 - src2; }
}

/* Operation >> */
OPERATION SHIFT_R {
   DECLARE {
      GROUP dest, src1, src2 = { ... }; }
   SYNTAX { dest "=" src1 ">>" src2 ";" }
   BEHAVIOR { dest = src1 >> src2; }
}

/* Control flow operations */
OPERATION IF_LT {
   DECLARE {
      GROUP src = { ... }; LABEL addr; }
   SYNTAX { "if(" src "< 0) goto" addr ";" }
   BEHAVIOR { if(src < 0 ) PC = addr-1; }
}

OPERATION GOTO { ... }
```

*Example 5.2:* Specification of a generic HLL model in LISA.

In order to execute the CORDIC application on the generic HLL model, the initial application code (cf. section 2.1) is revised to the one shown in example 5.3. The revision concerns the replacement of the division by $2^i$ by a right shift and the transformation of the control-flow into unconditional and conditional jumps. Besides, the variables X, Y, and Z are initialized depending on the value of the angle. In the example, the sine and cosine of an angle of 30 degrees is to be calculated and therefore the respective initial values are assigned.

The simulator derived from this LISA model constitutes a virtual machine executing the CORDIC application directly. The profiling capabilities of the simulator are used to generate execution statistics of the application code, where the cumulative percentage for the execution of each HLL statement becomes visible. Figure 5.3 shows the profiling results of the CORDIC application run on the simulator.

The cumulative percentage of execution of one line of the application code is indicated by the light gray bars in the application window. Besides, loops in the application code are identified and their total number of execution is displayed. To get an idea of the frequency of execution of operations, the total number of executions is shown as well.

```
/************************************/
/*  CORDIC application written      */
/*  in ANSI-C subset                */
/************************************/

/* Declaration of variables      */
/* takes place in LISA model     */

/* Initialize variables X,Y,Z    */
/* for calculation of sin/cos    */
/* of 30 degrees                 */

X: .word 60725332
Y: .word 0
Z: .word 34906585
```

```
/* Begin CORDIC calculation          */
_begin_loop: dx = X >> i;
             dy = Y >> i;
             dz = atan_lookup(i);
_begin_if:   if(Z < 0) goto _begin_else;
             X = X - dy;
             Y = Y + dx;
             Z = Z - dz;
             goto _end_loop;
_begin_else: X = X + dy;
             Y = Y - dx;
             Z = Z + dz;
_end_if:     cnt = i - N;
             i = i + 1;
_end_loop:   if(cnt < 0) goto _begin_loop;
/* End CORDIC calculation            */
```

*Example 5.3:* Target application to be executed on the generic HLL model.

In this architecture directly reflecting the application source code, the total number of control steps required to execute the CORDIC amounts to 104. Moreover, it can be seen that the loop within the application code has been executed ten times, which corresponds to the loop counter value $N$ in the underlying LISA model. Furthermore, as expected, the number of executions of the *if-else* branches is almost evenly distributed.

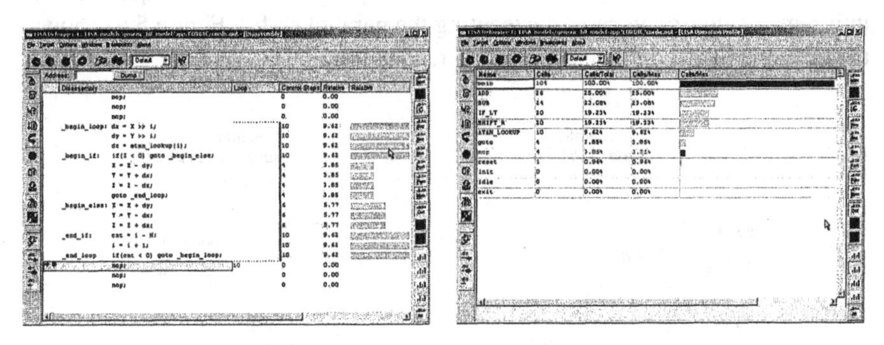

*Figure 5.3.* Application profiling in the LISA simulator.

As the application under test is very small, the information displayed by the LISA operation profiler is obvious: besides the reserved operation *main*, which drives the simulation and is thus executed in every control step, the addition and subtraction operations have been executed most often. As the basic blocks within the *if-else* branches are using purely arithmetic operations and since there is no data dependency among those operations, this is a potential point of optimization through parallelization to reduce cycle count. The same holds true for the two consecutive shift operations at the beginning of the CORDIC application.

In the next step of model refinement – the abstract resource model – such optimizations can be taken into consideration in the processor model by assigning operations in the LISA model to resources. By this, HLL-statements of the original application code can be executed in parallel.

## 2.3 Abstract Resource Model

Based on the profiling information gathered from running the application code on the generic HLL model, critical portions of the application code are identified that require parallelization. Besides, the operator usage and especially recurring orders of used operators indicate which operations rather should be realized with explicit hardware acceleration.

Therefore, the application code and the LISA processor model are revised to take parallel resources into account. It was shown in section 2.2 that the *if-else* branch with three subsequent arithmetic operations, which have no data dependency, are a potential candidate for parallelization. The equivalent hardware structure would be three parallel units capable of executing the addition and subtraction operations. Besides, the two shift operations at the beginning of the application code can be processed in parallel as well. To save hardware costs, these shift operations could be executed in the same units as the addition and subtraction in case of using an arithmetic logical unit (ALU). As for the basic blocks within the *if-else* branches three parallel units are required, one unit remains unused in case of executing the parallel shifts. Figure 5.4a shows a block diagram with the parallel ALU realization.

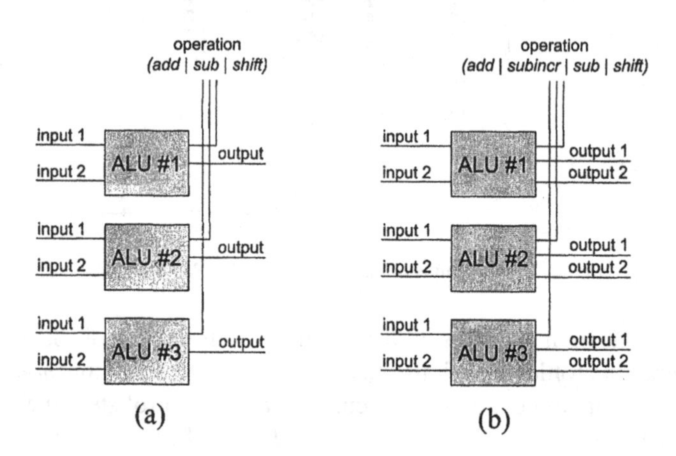


*Figure 5.4.* Parallel ALUs to decrease cycle-count in CORDIC application.

However, there is another potential point of optimization in the application code. The increment of the loop counter can be realized as a post-increment in the previous line of the application code without major additional hardware costs. Besides, the operation mirroring the decision if the loop is to be executed another time ($cnt = i - N$) can be moved in front of the loop as there is neither a dependency from the *if*- nor from the *else*-branch within the application code. As this operation mirrors again an arithmetic operation, it can take the free slot in one of the ALUs when executed in parallel with the two shift operations. The ALUs are extended to be capable of performing a post increment on the first input value (*input1*). Figure 5.4b shows the block diagram of the parallel ALU. The first output port carries the result of the ALU operation while the second contains the value of the incremented input value in case of executing the post increment operation.

```
/************************************/
/*  LISA model aware of parallel   */
/*  resources                      */
/************************************/

/* Declaration of variables */
RESOURCE { ...
}

/* Initialization of variables */
OPERATION reset {
  BEHAVIOR { ... }
}

/* Main operation */
OPERATION main {
  DECLARE {
    GROUP operations ={ PAR_ALU || ... }
  SYNTAX { operations }
  BEHAVIOR {
    operations();
    PC++;
  }
}

/* Parallel ALUs */
OPERATION PAR_ALU {
   DECLARE {
     GROUP ALU1, ALU2, ALU3 = {
        ADD || SUBINCR || SUB || SHIFT_R };
     }
   }
   SYNTAX { ALU1 "||" ALU2 "||" ALU3 }
   BEHAVIOR { ALU1(); ALU2(); ALU3(); }
}

/* Arithmetic and Logical Operations */
OPERATION ADD { ... }
OPERATION SUBINCR { ... }
OPERATION SUB { ... }
OPERATION SHIFT_R { ... }

/* Get ATAN Value */
OPERATION ATAN { ... }

/* Control Flow Operations */
OPERATION IF_LT { ... }
OPERATION GOTO { ... }
```

*Example 5.4:* Specification of an abstract resource model in LISA.

Example 5.4 shows the revised LISA code of the generic architecture model, which is now aware of parallel resources while example 5.5 shows the application to be executed on the model. In the LISA model, the new operation *PAR_ALU* is shown. It mirrors the ability of the hardware to execute certain operations simultaneously – *ADD*, *SUBINCR*, *SUB*, and *SHIFT_R*. The implementation of these operations in the LISA model is identical to the ones shown in example 5.2.

The application code is changed according to the syntax of parallel execution specified in the LISA model. In the example, the application remains unchanged compared to example 5.3 except for the logical *or* symbols (||) following the statements. The logical *or* symbols indicate that the next statement is being executed in parallel with the current.

```
/************************************/
/*  CORDIC application using        */
/*  parallel resources in model     */
/************************************/

/* Declaration of variables         */
/* takes place in LISA model        */
X: ..., Y: ..., Z: ...

/* Begin CORDIC calculation */
_begin_loop: dx = X >> i; ||
             dy = Y >> i; ||
             cnt = i++ - N;
             dz = atan_lookup(i);
_begin_if:   if(Z < 0) goto _begin_else;
             X = X - dy; ||
             Y = Y + dx; ||
             Z = Z - dz;
             goto _end_loop;
_begin_else: X = X + dy; ||
             Y = Y - dx; ||
             Z = Z + dz;
_end_loop:   if(cnt < 0) goto _begin_loop;

/* End CORDIC calculation */
```

*Example 5.5:* Target application to be executed on the parallelized generic HLL model.

As in the generic HLL model, the profiling capabilities of the simulator are used to gather execution statistics. Here it is examined, how the increasing ILP helps to cut down total execution time of the application running on the target architecture. The number of control steps required to execute the application program is decreased to 54. Therefore, in the example of the CORDIC application, the insertion of parallelism cuts down total execution time by approximately 50%. However, this comes at the cost of increased architectural complexity.

## 2.4 Data-Path Model

In the next phase of the processor design, the data-path of the processor is fixed. In addition to the parallelization of processor resources introduced in the abstract resource model, the algorithmic kernel functions are now transformed to the assembly level. As the data-path is working with physical processor resources like registers and memories, these need to be taken into account in the LISA model as well.

In this simple CORDIC example, the data-path is represented by the ALUs (cf. figure 5.4b) that allow processing three arithmetic and/or logical operations in parallel. Till now, only one of the design goals formulated at the beginning of this chapter has been addressed when optimizing the architecture: maximal performance. At this point in time, the second design goal can be taken into consideration as well, namely minimal program size. Minimal program size implies on the one hand high code density and on the other hand compact coding of instructions. Code density can only be examined and optimized once the micro-architecture is introduced. It refers to the compactness of the application which depends on the number of occurrences of *NOPs* (no-operations) in the application code due to a lack of ILP. With regard to compact instruction word coding, there is obviously a trade-off between programmability and width of the instruction word. The two extremes – full programmability of the architecture while maximizing instruction word width and minimal programmability while minimizing instruction word width – are illustrated in the following for the CORDIC application.

In the first case, all ALUs can work with arbitrary register resources as input and output values. In the application shown in example 5.5 there are at least nine different resources that need to be accepted by the ALU as input and output. Besides, each of the ALUs can process addition, subtraction with post-increment, substraction, and shifting. In the latter case, the instruction word width can be minimized by binding each operation in each of the ALU to special processor resources. However, this minimizes programmability and thus limits the usage of the resulting highly specialized instruction to the kernel of CORDIC angle calculation.

The resulting formation of the instruction word for both eventualities is depicted in figure 5.5.

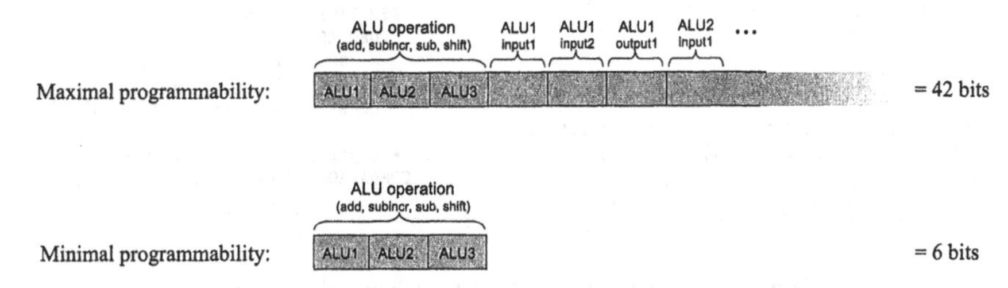


*Figure 5.5.* Trade-off between instruction word width and programmability.

As there are four alternative operations possible for each ALU, the opcode for each ALU amounts to two bits. Furthermore, to code one resource out of at least nine possible, four bits are required for each of the input and output ports. Therefore, the total number of bits required for the instruction providing maximal programmability is 42. For the second case, where all operations are hardwired with special resources and which thus provides minimal programmability, the instruction word is compressed to six bits. As in this sample exploration minimum program size is a design goal, the latter solution is chosen. It is obvious that the resulting special-purpose instruction is highly application-specific, as it is working with fixed processor resources for the respective ALU operations. However, this is a typical case in ASIPs that contain highly application-specific instructions to cope with diametrical requirements like parallelism demanding for very long instruction words (VLIW) and minimum program size. Example 5.6 shows the revised LISA description of the architecture.

Compared to the previous processor design phases, the LISA model now works with physical resources. The resource section declares a register bank *R* consisting of nine elements and a data memory *data_rom*. The parallel executed

```
/*******************************************/
/*  LISA model with instructions           */
/*  for data-path and processor            */
/*  resources                              */
/*******************************************/

/* Declaration of processor resources */
RESOURCE {
  PROGRAM_COUNTER int PC;
  REGISTER int R[0..8];
  DATA_MEMORY data_rom[0..0xf];
  ...
}

/* Initialization of processor resources */
OPERATION reset { ... }
/* Main operation */
OPERATION main {
  DECLARE {
    GROUP operations ={ PAR_ALU || ... }
  SYNTAX { operations }
  BEHAVIOR { operations(); PC++; }
}

/* Parallel ALUs */
OPERATION PAR_ALU {
   DECLARE {
      INSTANCE ALU1,ALU2,ALU3;
   }
   SYNTAX { "PAR_ALU" ALU1 "," ALU2 "," ALU3 }
   BEHAVIOR { ALU1(); ALU2(); ALU3(); }
}
 OPERATION GOTO { ... }
```

```
/* Operations in ALU1 */
OPERATION ALU1 {
   DECLARE { ENUM op =
      ADD, SUBINCR, SUB, SHIFT_R;
   }
 SWITCH(op) {
  CASE ADD: {
    SYNTAX { "add" }
    BEHAVIOR { R[1]+=R[5];}}
  CASE SUBINCR: {
    SYNTAX { "subi" }
    BEHAVIOR { R[7]=R[0]++-R[8];}}
  CASE SUB: {
    SYNTAX { "sub" }
    BEHAVIOR { R[1]-=R[5];}}
  CASE SHIFT_R: {
    SYNTAX { "shr" }
    BEHAVIOR { R[4]=R[1]>>R[0];}}}}

/* Operations in ALU2 + ALU3 */
OPERATION ALU2 { ... }
OPERATION ALU2 { ... }

/* Get Value from Memory */
OPERATION LOAD_MEM {
   DECLARE {
      GROUP dest = { register }; }
   SYNTAX { "LD_MEM" dest }
   BEHAVIOR {dest=data_rom[R[0]];}}

/* Control Flow Operations */
OPERATION IF_LT { ... }
```

*Example 5.6:* Specification of a data-path model in LISA.

arithmetic/logical operations of the application code are now summarized in one instruction *PAR_ALU*. This instruction is parameterized by the operations to be executed concurrently on each of the three ALUs – *add*, *subi*, *sub*, and *shr*. As explained, the registers used in context of each operation on the respective ALUs are hardwired and cannot be changed. Register *R[0]* contains the application loop counter *i*, while registers *R[1]* to *R[6]* imply the variables *X*, *Y*, *Zdx*, *dy*, and *dz* of the application C-code respectively. The arctan values are stored in a lookup table in the data memory *data_rom*. Using the *LD_MEM* instruction of the architecture, they can be loaded into an arbitrary register.

The revised application code is shown in example 5.7. The HLL statements concatenated by the logical *or* symbol indicating parallel execution (cf. example 5.5) are replaced by the assembly code of the special-purpose instruction *PAR_ALU*. Moreover, all variable names are replaced by physical resources. Except for the control-flow, which is still formulated in the original C-style, the complete application is now formulated in assembly code.

At this point in time, the resource profiler can be used to examine the resource utilization during the application run. Obviously, in this simple CORDIC example making use of only few processor resources, the resource profiler is not

```
/**************************************/
/*  CORDIC application using          */
/*  assembly language for kernel      */
/**************************************/

/* Declaration of variables           */
/* takes place in                     */
/* LISA model                         */

/* Initialization of resources for    */
/* calculation of sin/cos of 30       */
R1: ..., R2: ..., R3: ...
```

```
/* Begin CORDIC calculation */
_begin_loop: PAR_ALU shr, shr, subi
             LD_MEM R6
_begin_if:   if(R3 < 0) goto _begin_else;
             PAR_ALU add,sub,add
             goto _end_loop;
_begin_else: PAR_ALU sub,add,sub
_end_loop:   if(R7 < 0) goto _begin_loop;

/* End CORDIC calculation */
```

*Example 5.7:* Application with algorithmic kernel functions realized in assembly code.

necessarily required as the utilization of resources is obvious anyway. However, in larger assembly programs and more complex architectures this does not hold true. In this case, the developer cannot keep the overview on the resource utilization without tool support. Furthermore, the resource profiler is most valuable in cadence with using a HLL C-compiler. Here, the application is not changed during the architecture exploration phase. Instead, the retargetable compiler is used to generate automatically the most efficient executable program for the given target processor architecture. Thereby, the resource profiler points the architecture designer on the one hand to resources that are not used by the compiler and can thus be removed from the architecture and on the other hand to possible congestions resulting in suboptimal code quality that require additional resources to resolve.

The profiler generated from LISA processor models displays information on the number of reads and writes to processor resources during the application run. Moreover, the number of simultaneous accesses to a group of resources, like e.g. a register bank or a memory, are indicated. Figure 5.6 shows the resource profiling results for the CORDIC application running on the architecture model worked out in this section.

As this phase in the architecture exploration process is mainly concerned with the introduction of the assembly syntax and the processor resources to the model, the number of control steps required to execute the application on the target architecture remains the same as in the previous phase. Optimization of the cycle count is addressed again in the next phase of the model refinement process – the *instruction-based model.*

## 2.5 Instruction-Based Model

After the data-path (i.e. the processing intensive algorithmic kernels) is optimized and the respective segments of the application code are transformed to the assembly level, the remaining part of the application is addressed. For this purpose, more instructions are introduced which can be characterized as being

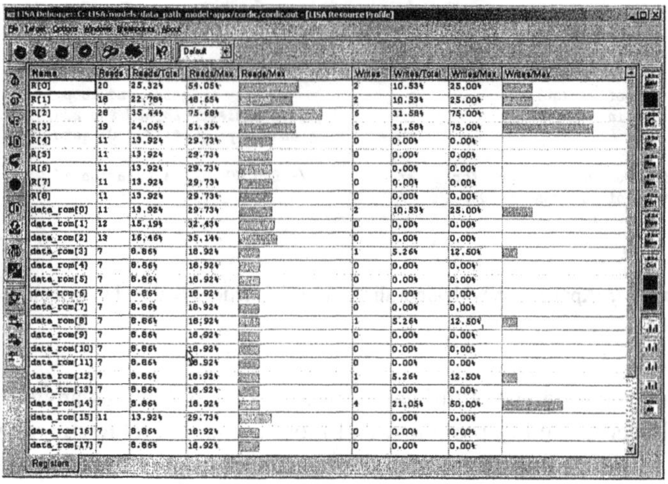

| Name | Reads | Reads/Total | Reads/Max | Reads/Max | Writes | Writes/Total | Writes/Max | Writes/Max |
|---|---|---|---|---|---|---|---|---|
| R[0] | 20 | 25.32% | 54.05% | | 2 | 10.53% | 25.00% | |
| R[1] | 18 | 22.78% | 48.65% | | 2 | 10.53% | 25.00% | |
| R[2] | 28 | 35.44% | 75.68% | | 6 | 31.58% | 75.00% | |
| R[3] | 19 | 24.05% | 51.35% | | 6 | 31.58% | 75.00% | |
| R[4] | 11 | 13.92% | 29.73% | | 0 | 0.00% | 0.00% | |
| R[5] | 11 | 13.92% | 29.73% | | 0 | 0.00% | 0.00% | |
| R[6] | 11 | 13.92% | 29.73% | | 0 | 0.00% | 0.00% | |
| R[7] | 11 | 13.92% | 29.73% | | 0 | 0.00% | 0.00% | |
| R[8] | 11 | 13.92% | 29.73% | | 0 | 0.00% | 0.00% | |
| data_rom[0] | 11 | 13.92% | 29.73% | | 2 | 10.53% | 25.00% | |
| data_rom[1] | 12 | 15.19% | 32.43% | | 0 | 0.00% | 0.00% | |
| data_rom[2] | 13 | 16.46% | 35.14% | | 0 | 0.00% | 0.00% | |
| data_rom[3] | 7 | 8.86% | 18.92% | | 1 | 5.26% | 12.50% | |
| data_rom[4] | 7 | 8.86% | 18.92% | | 0 | 0.00% | 0.00% | |
| data_rom[5] | 7 | 8.86% | 18.92% | | 0 | 0.00% | 0.00% | |
| data_rom[6] | 7 | 8.86% | 18.92% | | 0 | 0.00% | 0.00% | |
| data_rom[7] | 7 | 8.86% | 18.92% | | 0 | 0.00% | 0.00% | |
| data_rom[8] | 7 | 8.86% | 18.92% | | 1 | 5.26% | 12.50% | |
| data_rom[9] | 7 | 8.86% | 18.92% | | 0 | 0.00% | 0.00% | |
| data_rom[10] | 7 | 8.86% | 18.92% | | 0 | 0.00% | 0.00% | |
| data_rom[11] | 7 | 8.86% | 18.92% | | 0 | 0.00% | 0.00% | |
| data_rom[12] | 7 | 8.86% | 18.92% | | 1 | 5.26% | 12.50% | |
| data_rom[13] | 7 | 8.86% | 18.92% | | 0 | 0.00% | 0.00% | |
| data_rom[14] | 7 | 8.86% | 18.92% | | 4 | 21.05% | 50.00% | |
| data_rom[15] | 11 | 13.92% | 29.73% | | 0 | 0.00% | 0.00% | |
| data_rom[16] | 7 | 8.86% | 18.92% | | 0 | 0.00% | 0.00% | |
| data_rom[17] | 7 | 8.86% | 18.92% | | 0 | 0.00% | 0.00% | |

*Figure 5.6.* Resource profiling in the LISA debugger.

of *general-purpose* nature. This includes instructions handling the control-flow in the architecture. Furthermore, an instruction decoder is added to the model which requires at first the specification of the binary coding for each instruction in the model.

In this sample exploration, there is no need to introduce further instructions to the model except for instructions handling the control-flow in the architecture. For the *if-else* construct, simple conditional (*JMP_LT*) and unconditional (*JMP*) jump instructions are introduced that branch to an arbitrary address that is specified as a parameter. One possibility of optimization in the target architecture executing the CORDIC application is the loop that embraces the complete application kernel. After each iteration, a condition is checked to determine if the loop is to be repeated or not. The enhancement concerns the introduction of a zero overhead loop (ZOLP) to the architecture. Here, an instruction is required that initializes a loop counter that is decremented automatically in hardware at the end of each iteration. Besides, the ZOLP hardware checks automatically if the program-counter equals the end of the loop to branch back to the beginning. For this purpose, the *ZOLP* instruction is parameterized with the loop count and the end address of the loop. The begin of the loop is implicitly fixed by the instruction following the *ZOLP* instruction.

Example 5.8 shows an excerpt of the LISA model which now contains all instructions in the architecture with their binary coding, assembly syntax, and behavior. As the processor model is now aware of all instructions and their respective binary coding, the resource section of the LISA model is extended by the declaration of a program memory (*prog_rom*). Based on the information

```
/****************************************/
/*  LISA model with all instructions    */
/*  including binary coding, assembly   */
/*  syntax and behavior                 */
/****************************************/

/* Declaration of processor resources */
RESOURCE {
  PROGRAM_COUNTER int PC;
  REGISTER int R[0..8];
  DATA_MEMORY       data_rom[0..0xf];
  PROGRAM_MEMORY    prog_rom[0..0xff];
  int insn_reg;
}

/* Initialization of processor resources */
OPERATION reset {
  BEHAVIOR {  ... }
}

/* Main operation */
OPERATION main {
  BEHAVIOR {
    insn_reg = prog_mem[PC];
    Decode_Instruction(); PC++;
  }
}

/* Operation Decode */
OPERATION Decode_Instruction {
   DECLARE {
      GROUP operation = { PAR_ALU || .. }
   }
   CODING { insn_reg == operation }
   SYNTAX { operation }
   BEHAVIOR { operation(); }
}
```

```
/* Parallel ALUs */
OPERATION PAR_ALU {
   DECLARE {
     INSTANCE ALU1,ALU2,ALU3;
   }
   CODING { 0b00 ALU1 ALU2 ALU3 0bx[2] }
   SYNTAX { "PAR_ALU" ALU1 "," ALU2 "," ALU3 }
   BEHAVIOR { ALU1(); ALU2(); ALU3(); }
}

/* Operations in ALU1 */
OPERATION ALU1 {
   DECLARE { ENUM alu_op =
     ADD, SUBINCR, SUB, SHIFT_R;
   }
   SWITCH(alu_op) {
     CASE ADD: { CODING { 0b00 } SYNTAX ... }
       ...
     }
   }

/* Operation ZOLP */
OPERATION ZOLP {
   DECLARE {
     LABEL addr, lpcnt;
   }
   CODING { 0b01 lpcnt=0bx[4] addr=0bx[4] }
   SYNTAX { "ZOLP" lpcnt=#U SYMBOL(addr=#U) }
   BEHAVIOR { ... }
}

/* Other Operations */
OPERATION ALU2 { ... }
OPERATION ALU3 { ... }
OPERATION JMP { ... }
OPERATION JMP_LT { ... }
...
```

*Example 5.8:* Specification of an instruction-based model in LISA.

on the binary coding of the instructions, the code generation tools generated from the LISA description – assembler and linker (cf. chapter 7.1) – generate a binary executable object file which is used to initialize the program memory at the start-up of the simulator.

At simulation run-time, each instruction to be executed is first fetched from memory and loaded into the instruction register *insn_reg*. Based on the content of this register, the operation *Decode_Instruction* picks the valid operations to be executed in the context of the current instruction. Example 5.9 shows the optimized assembly code of the CORDIC application for the target architecture.

In order to benchmark the architecture for its conformance with the design goals, the instruction-set profiler is utilized. Figure 5.7 shows the results of the instruction profiler when running the application on the target architecture. The number of control steps required to execute the application has been reduced to 44. Obviously, it is strongly dependent on the micro-architecture implementation if the respective instructions can be executed within one cycle or not (cf. section 2.6).

```
/***********************************/
/*  CORDIC application using      */
/*  assembly language for         */
/*  complete application          */
/***********************************/

/* Declaration                     */
/* of resources takes place in     */
/* LISA model                      */

/* Initialization of resources for */
/* calculation of sin/cos of 30    */
R1: ..., R2: ..., R3: ...

/* Begin CORDIC calculation */
              ZOLP 10,_end_loop
_begin_loop:  PAR_ALU shr, shr, subi
              LD_MEM R6
_begin_if:    JMPLT R3, _begin_else
              PAR_ALU add,sub,add
              JMP _end_loop
_begin_else:  PAR_ALU sub,add,sub
_end_loop:

/* End CORDIC calculation */
```

*Example 5.9:* Assembly code of target application using the specified instruction-set.

Besides, the total program size, which depends on both the instruction word width and the number of instructions in the program code has been minimized. The instruction word width required to code the instructions of the sample architecture amounts to eight bits. As explained in section 2.4, this is only

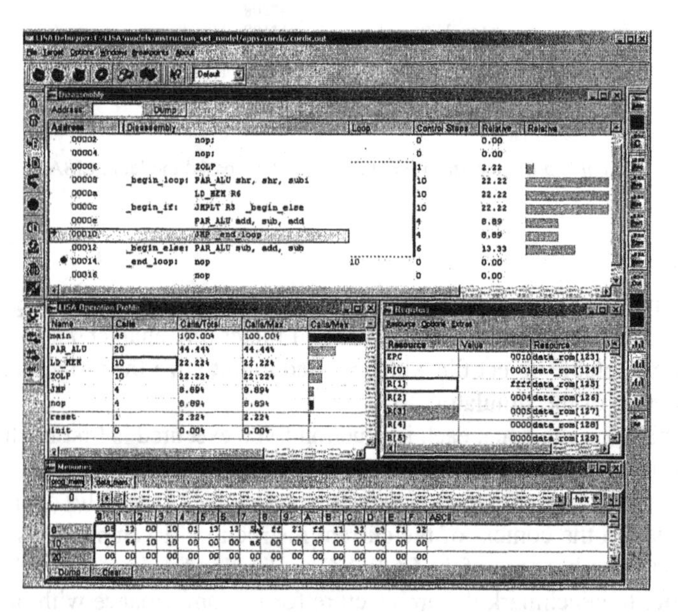

*Figure 5.7.* Instruction profiling in the LISA simulator.

possible because of the parallel ALUs working with fixed registers, which thus do not have to be coded in the instruction word. This comes at the cost of limited programmability. The total size of the program code consisting of a total of seven instructions amounts to 56 bits.

## 2.6 Cycle-Based Model

The next phase in the processor design flow is concerned with the refinement of the LISA model from instruction-based to cycle-based. For this purpose the model is enriched by architectural information on pipelines as well as the distribution of instructions' execution over several cycles and pipeline stages. Pipelines are used in processor architectures to hide the latency between the execution of instructions.

Obviously, the specification and optimization of the architecture is strongly dependent on the next phase in the processor design process: the introduction of the RTL micro-architectural model. Only models on this level can be used to generate synthesizable HDL code (cf. chapter 6) which is required to realize a gate-level model and to evaluate the architecture for power consumption and maximum clock frequency. Naturally, this information strongly influences the appearance of e.g. the pipeline. In this sample exploration, the synthesis results are not taken into account as they are extensively presented for a similar *real-world* architecture in chapter 6. For the sample processor resulting from the refinement process in the previous sections executing the CORDIC application, a simple three stage pipeline has been chosen. Figure 5.8 shows a pictural

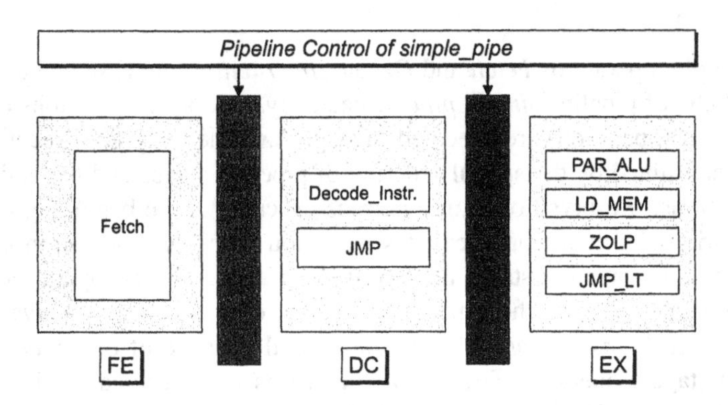


*Figure 5.8.* Micro-architecture of processor executing the CORDIC application.

representation of the resulting architecture and the assignment of LISA operations to the respective pipeline stages. Example 5.10 displays the respective LISA processor model. The resource section in the model is enhanced by the definition of the pipeline which is composed of three stages – fetch (FE), decode (DC), and execute (EX).

According to the specification of the timing model in LISA (cf. chapter 3.2.5), operations are assigned to those pipeline stages they are executed in. In

```
/************************************/
/*  LISA model with instruction     */
/*  distribution over several       */
/*  pipeline stages                 */
/************************************/

/* Declaration of processor resources */
RESOURCE { ...
  PIPELINE simple_pipe = { FE; DC; EX };
}

/* Initialization of resources */
OPERATION reset { ... }

/* Main operation */
OPERATION main {
  DECLARE { INSTANCE Fetch; }
  ACTIVATION { Fetch }
  BEHAVIOR {
    PIPELINE(simple_pipe).execute();
    PIPELINE(simple_pipe).shift();
  }
}

/* Operation Fetch */
OPERATION Fetch IN simple_pipe.FE {
  DECLARE { INSTANCE Decode; }
  ACTIVATION { Decode }
  BEHAVIOR { insn_reg = prog_mem[PC]; }
}

/* Operation Decode */
OPERATION Decode IN simple_pipe.DC {
  DECLARE {
     GROUP operation = { PAR_ALU || ... }}
  CODING { insn_reg == operation }
  SYNTAX { operation }
  ACTIVATION { operations }
}

/* Unconditional Jump Instruction */
OPERATION JMP IN simple_pipe.DC { ... }

/* Conditional Jump Instruction */
OPERATION JMP_LT IN simple_pipe.EX {
   DECLARE { LABEL addr; }
  CODING { ... } SYNTAX { ... }
  BEHAVIOR {
    if(cond_reg < 0) {
      PC = addr;
      PIPELINE(simple_pipe,FE/DC).flush();
      PIPELINE(simple_pipe,DC/EX).flush();
      PC++;
    }
} }

/* Other operations */
OPERATION PAR_ALU IN simple_pipe.EX { ... }
OPERATION LD_MEM IN simple_pipe.EX { ... }
OPERATION ZOLP IN simple_pipe.EX { ... }
```

*Example 5.10:* Specification of a cycle-based model in LISA.

the example, operations *Fetch* and *Decode_Instructions* are assigned to stages *FE* and *DC* of pipeline *simple_pipe* respectively. All other operations (except for the *JMP* operation) are executed in stage *EX*. The temporal execution order is determined by the spatial ordering of pipeline stages and the activation of operations. Reserved operation *main* is executed at the beginning of each cycle during simulation and activates operation *Fetch*. As operation *main* is assigned to no pipeline stage, the execution of the activated operation takes place immediately. At the time of execution, operation *Fetch* activates operation *Decode_Instruction*. Due to the spacial distance of one between the pipeline stages, operation *Decode_Instruction* is not executed until the next cycle. Subsequently, operation *Decode_Instruction* activates one of the operations assigned to the *EX* stage, depending on the coding of the instruction word. Using this flexible mechanism, an operation activation chain is built up.

The activation of operations can be seen as setting control signals in hardware, which are forwarded through the pipeline with each rising clock edge. It will be shown in chapter 6 that the information on the processor control is in fact extracted from the activation mechanism in conjunction with the instruction word coding. The execution of activated operations and the shifting of the pipeline is performed in the behavior section of operation *main* using the predefined pipeline operations *shift* and *execute*, which are provided by the LISA environment.

The jump instructions deserve special attention: the unconditional branch is executed in the decode stage (*DC*) and is followed by a delay slot. This means that the instruction following the *JMP* instruction is always executed. In case there is not enough ILP in the application code, the delay slot needs to be filled with a *NOP* instruction. On the contrary, the conditional jump instruction *JMP_LT* is executed in the *EX* stage of the pipeline, i.e. the branch target instruction is not executed until in two cycles. However, there are already two instructions in the pipeline that follow the branch instructions but must not be executed. This conditional jump contains no delay slot. Therefore, the *JMP_LT* instruction flushes stages *FE* and *DC* in case the branch is taken by using the pipeline control function *flush()*, which is provided by the generic machine model of the LISA environment.

```
/**********************************/
/*  CORDIC application using      */
/*  assembly language for         */
/*  complete application          */
/**********************************/

/* Declaration                    */
/* of resources takes place in    */
/* LISA model                     */

/* Initialization of resources for*/
/* calculation of sin/cos of 30   */
R1: ..., R2: ..., R3: ...
```

```
/* Begin CORDIC calculation */
              ZOLP 10,_end_loop
_begin_loop:  PAR_ALU shr, shr, subi
              LD_MEM R6
_begin_if:    JMPLT R3, _begin_else
              JMP _end_loop
              PAR_ALU add,sub,add
_begin_else:  PAR_ALU sub,add,sub
_end_loop:

/* End CORDIC calculation */
```

*Example 5.11:* Assembly code of the CORDIC under consideration of pipeline effects.

Example 5.11 shows the revised CORDIC application which takes pipelining effects into account. The application has not changed significantly compared to the application optimized for the instruction-based model (cf. example 5.9) – except for the placement of the unconditional branch instruction in the application code. As explained, this instruction is followed by a delay slot which is filled with the *PAR_ALU* by switching the ordering of those instructions in the application code.

Especially in more complex architectures as this one, in order to optimize the architecture in cadence with the application, the pipeline profiling capabilities of the LISA simulator can be used. Figure 5.9 shows the pipeline profiler provided by the LISA environment. The pipeline profiling window displays information about the operations in the pipeline, stalls or flushes issued, and statistics on how many times a stall or flush was issued by a certain stage. The disassembly window shows with a gray bar which instructions caused hazards and how many times. Based on this information, the architecture is refined in an iterative process until it meets the posed design criteria. In the sample

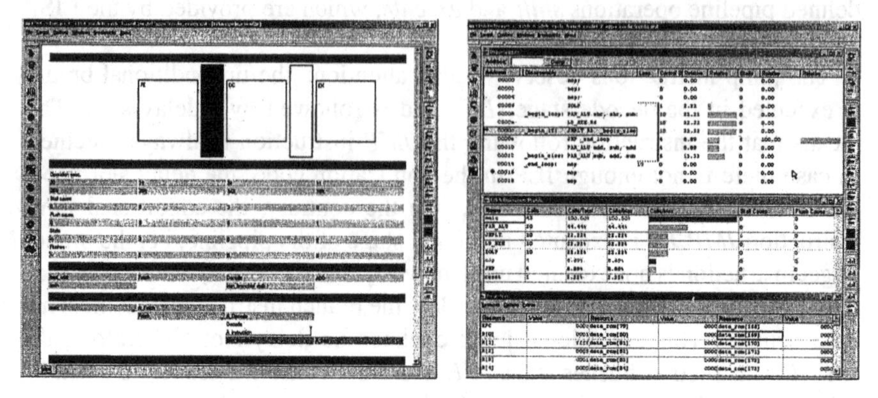

*Figure 5.9.* Pipeline profiling in the LISA simulator.

architecture, the size of the program has not changed compared to the previous chapter. However, due to the issuing of flushes to the *FE* and *DC* stages in context of the conditional branch instruction, which delays the execution of the branch target instruction by two cycles, the total number of control steps increases to 56. Obviously, on this level of abstraction, control steps are equal to cycles. However, the design goal – to maximize performance – is dependent on both cycle count and clock frequency. Clock frequency can be increased by pipelining to reduce latency between the execution of instructions.

## 2.7 RTL Micro-Architectural Model

In the final step of the architecture exploration phase, the architecture is refined to the level of register transfers. On this level, the HDL code generator can be used (cf. chapter 6) to gather meaningful numbers on hardware cost and performance parameters, e.g. design size, power consumption, and clock frequency. As indicated in chapter 3.2.6, the sample architecture developed within the scope of this chapter is not taken through synthesis as this step is discussed in detail in chapter 6. The motivation to show this additional refinement step is to indicate how the cycle-based model has to be enhanced to come to a model that is capable of HDL-code generation and thus synthesis.

Example 5.12 displays the refined LISA model of the target architecture optimized for running the CORDIC application. The differences to the cycle-based, pipelined model from the previous design phase concern the usage of pipeline registers to shift values through the pipeline and the explicit setting of data values by operations. This additional information is required to enable HDL-code generation from LISA processor models.

This concludes the LISA-based processor design flow by exploring and optimizing the architecture based on profiling and implementation results. It is

```
/************************************/
/*  LISA model with micro-          */
/*  architectural information on    */
/*  pipeline registers, etc.        */
/************************************/

/* Declaration of */
RESOURCE {
  PIPELINE simple_pipe = { FE, DC, EX };
  PIPELINE_REGISTER IN simple_pipe {
     REGISTER int insn_reg;
     bit[32] operand A;
     bit[32] operand B;
  }
}

/* Initialization */
OPERATION reset { ... }

/* Main operation */
OPERATION main { ... }

/* Other operations */ ...
```

```
/* Operation Fetch */
OPERATION Fetch IN simple_pipe.FE {
   DECLARE { INSTANCE Decode; }
   ACTIVATION { Decode }
   BEHAVIOR USES ( IN prog_mem[]; ) {
     PIPELINE_REGISTER(simple_pipe,FE/DC)
     .insn_reg = prog_mem[PC];
   }
}

/* Operation Decode */
OPERATION Decode IN simple_pipe.DC {
   DECLARE { GROUP operation = ...;
             GROUP src ... ;
   }
   CODING {PIPELINE_REGISTER(simple_pipe,FE/DC)
           .insn_reg == operation src }
   SYNTAX { operation }
   ACTIVATION { operation }
   BEHAVIOR SET (
      PIPELINE_REGISTER(simple_pipe,DC/EX)
      .operandA = src; )
   }
```

*Example 5.12:* Specification of an RTL micro-architectural model in LISA.

obvious that multiple iterations on each level of abstraction of the presented processor design process are required to optimize the architecture for the target application. Naturally, these iterations are not only limited to a single phase but can stretch over several phases.

## 3. Concluding Remarks

In this chapter, the first quadrant of the LISA processor design platform was introduced – architecture exploration. It was shown that ASIP design using LISA can be performed by seamlessly refining a LISA model of the target architecture from an application-centric level to the level of the micro-architectural implementation. For this purpose, six phases within the processor design process were identified that allow optimizing the processor on different levels of abstraction. For each step during the processor design process, a LISA model can be realized and a working set of software development tools can be generated automatically. The profiling capabilities of the tool-suite can be used to evaluate the architecture and to optimize the application in cadence with the architecture on the respective level of abstraction.

Obviously, it is required to take data on maximum clock frequency, design size, and power consumption of the target architecture into account as well. Therefore, the step to implementation has to be included into the exploration loop. Architecture implementation from LISA processor models is discussed in the next chapter.

# Chapter 6

# ARCHITECTURE IMPLEMENTATION

The outcome of the architecture exploration phase is an architectural model at the level of register transfer (RT) accuracy. This model is appropriate to optimize and benchmark the architecture in terms of throughput and resource requirements. However, to get meaningful data on hardware cost and performance parameters (e.g. design size, power consumption, clock frequency) the processor model has to be taken through the standard synthesis flow. By definition, the LISA language targets the description of programmable architectures in terms of their instruction-set, behavior, and structure. Detailed hardware structures are not modeled for the complete architecture as this would contradict the idea of generating fast instruction-set simulators (cf. chapter 2.1). On the level of RT accurate LISA models this concerns especially the description of the instruction decoding procedure, processor control, and behavioral code which is still carried out in an abstract way.

Today's standard synthesis tools [30, 131] work with models in hardware description languages (HDL) like VHDL and Verilog, though. Therefore, the ISA description in LISA resulting from the architecture exploration phase has to be manually transformed into synthesizable HDL code. This marks a break in the design flow as the processor model has to be re-implemented using a different specification language. It is obvious that this proceeding has significant disadvantages:

- re-implementing the processor in an HDL is a tedious and lengthy process that is prone to errors,
- the models need to be verified against each other, and
- two models of the target architecture need to be maintained by the designer.

For the above mentioned reasons, an automated solution is desirable which cuts down these disadvantages.

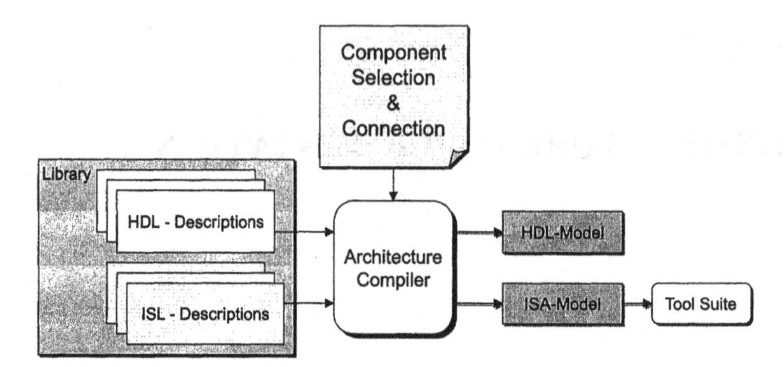


*Figure 6.1.* Module library approach to generate HDL code.

There are two ways of addressing this problem. Firstly, a library of predefined modules can be provided with the environment. This approach is also known as the *module library approach* [81]. The principle is depicted in figure 6.1. Here, a processor template library contains a set of pre-defined, highly optimized modules. Several representations exist for each module: on the one hand, the modules are available as pure C/C++ models wrapped with ISA information on binary coding, syntax, and timing as required for building the software development tools. In [47] it is reported that among these ISA modules the designer can even choose between different levels of abstraction. On the other hand, for each module there is also synthesizable HDL code available.

The architectural model is now built by selecting components from the library and by interconnecting those. This information is processed by an architecture compiler which generates both the complete HDL model of the target architecture and an ISA description which can be used for the generation of software development tools. The advantage of this approach is at hand: as the user operates with a set of predefined, highly tailored components, the quality of the generated HDL code is comparable to that of a hand-optimized version.

However, the module library approach has one significant drawback that excludes its usage in the design of application-specific processors. The module library approach works with a limited set of modules that are available to the processor designer. However, in ASIP design, every architecture is unique to the target application with highly tailored control and data-path for the dedicated application. The module library approach is suitable for architecture construction sets that reconfigure a base architecture, as e.g. companies like Tensilica [117] or ARC [144] do.

The second approach, which was chosen in the LISA processor design platform, is to generate the synthesizable HDL code of the target architecture from the ISA model [145, 146].

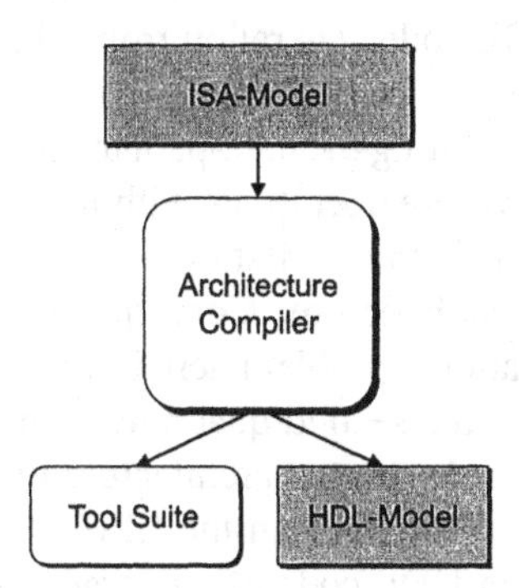


*Figure 6.2.* Generation of software tools and HDL model from ISA description.

The transformation process is shown in figure 6.2. However, even on the level of RT micro-architectural accuracy, the ISA model is more abstract than the synthesizable HDL model (cf. chapter 3.2.7). As the LISA processor design platform is targeting the development of ASIPs, which are highly optimized for one specific application domain, the HDL code generated from a LISA processor description has to fulfill tight constraints to be an acceptable replacement for HDL code handwritten by experienced designers.

For this reason, a semi-automatic approach was chosen, which distinguishes between those parts of the architecture that can be generated from the more abstract ISA specification and those that need to be specified manually. The portion of the architecture which needs to be specified twice – on the one hand in the ISA description and on the other hand in the HDL model – concerns the data-path within functional units. Using an image, this part of the architecture could be referred to as the *muscles*. As the data-path is crucial in terms of the critical path, power, and size, it must in most cases be optimized manually. Frequently, full-custom design technique must be used to meet power consumption and clock speed constraints.

For this reason, the generated HDL code is focusing on the following parts of the architecture:

- coarse processor structure such as register-set, pipeline, and pipeline registers,
- instruction decoder setting data and control signals which are carried through the pipeline and activate the functional units executed in context of the decoded instruction, and

- pipeline controller handling different pipeline interlocks, pipeline register flushes and supporting mechanisms such as data forwarding.

Again, using an image the above mentioned parts of the architecture which are considered for the HDL code generation from LISA could be referred to as the *brain* of the processor.

The disadvantage of re-writing the data-path in the HDL description by hand is that the behavior of hardware operations within those functional units has to be described and maintained twice – first in the LISA model and second in the HDL model of the target architecture. Consequently, in addition to the effort spent on each design iteration a problem here is verification (cf. chapter 9).

This chapter introduces the second quadrant of the LISA processor design platform which is concerned with the architecture implementation. It will be shown, how different information from the LISA description can be used to generate the corresponding HDL code of the target architecture. To prove the concept, a case study is presented that was carried out in cooperation with Infineon Technologies. Here, an ASIP for digital video broadcast terrestrial (DVB-T) was developed which was completely designed using the LISA methodology and tools – the ICORE. Area, power consumption, and the speed of the ICORE are compared to a hand-optimized version that was manually realized in previous work [124].

## 1. The ICORE Architecture

The ICORE (ISS[1]-CORE) is a low-power ASIP for DVB-T acquisition and tracking algorithms. For this reason, the ICORE is also referred to as DVB-T Post Processing Unit (PPU). The tasks of this architecture are the FFT-window-positioning, sampling-clock synchronization for interpolation/decimation, and carrier frequency offset estimation. The most significant feature of the architecture are highly optimized instructions supporting a fast CORDIC angle calculation (cf. chapter 5.2). Besides these special-purpose instructions, the ICORE is composed of 60 RISC-like instructions. Further information about the micro-architecture including pipelining will be presented in the following sections.

## 2. Architecture Generation from LISA

According to chapter 3.3, LISA allows the specification of models of the target architecture on different levels of accuracy. These abstraction levels vary from the level of a virtual machine directly executing the algorithmic kernel in C (cf. chapter 5.2.2) to the level of micro-architectural implementation (cf.

[1]Here, ISS denotes the abbreviation of the Institute for Integrated Signal Processing Systems.

chapter 5.2.7), where the latter marks the lowest abstraction level supported by the LISA language.

However, the abstraction level of an implementation model in a hardware description language like VHDL or Verilog, that can be processed by synthesis tools, is still below that. Figure 6.3 shows several levels of abstraction versus model complexity.

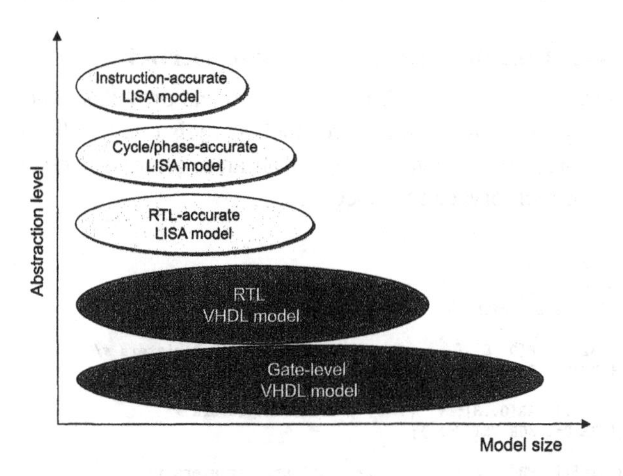


*Figure 6.3.* Model abstraction levels and specification languages in processor design.

It is obvious, that generating the HDL model from a LISA description requires the extraction of different architectural information which is captured in different processor models (cf. chapter 3.2) in LISA. The following model information is used to generate equivalent HDL code:

- the *memory model* provides information about the structure like registers, pipelines, pipeline registers, and memories,
- the *resource model* determines the number of data and address buses to be generated to access an arrangement of processor resources like registers or memories,
- the *instruction-set model* and the *timing model* provide the required information to generate decoders in hardware setting data and control signals as well as the pipeline controller, and
- the *behavioral model* and the *micro-architectural model* enable grouping of operations to functional units, the explicit setting of values to processor resources by decoders, and the generation of interconnections to processor resources.

Since the HDL model partly contains more structural information on the target architecture than the underlying LISA model, presumptions need to be made on the design of architectural elements, like e.g. the register-file implementation. Besides, information that is implicitly contained in the LISA model like exact timing and concurrency of hardware operations needs to be extracted.

The following sections are concerned with the way of extracting information on different parts of the processor architecture from LISA machine descriptions.

## 2.1 Extraction of Course Processor Structure

The resource section in LISA processor models provides general information on the structure of an architecture, such as registers, pipelines, buses, and memories. Moreover, the availability of an arrangement of resources, e.g. a register bank or memory, can be specified.

```
RESOURCE {
  MEMORY_MAP {
      0x000 -> 0x7ff, BYTES(4) : ROM[0..7ff], BYTES(4);
      0x800 -> 0x9ff, BYTES(4) : RAM[0..1ff], BYTES(4);
  }
  REGISTER S32      R[0..8] {            /* General Purpose Registers */
     PORT { READ=6 OR WRITE=6 };
  };
  REGISTER bit[11] AR[0..3]; {           /* Address Registers */
     PORT { READ=1 XOR WRITE=1 };
  };
  REGISTER bit[1]  PR[0..3]; {           /* Predicate Registers */
     PORT { READ=1 XOR WRITE=1 };
  };
  REGISTER bit[1]  SR[0..2];             /* Status Registers */

  REGISTER bit[1] zolp_state;
  REGISTER U32     zolp_curr_count;      /* Counters and Address Registers for Zero */
  REGISTER U32     zolp_end_count;       /* Overhead Loops: */

  DATA_MEMORY      S32 RAM[0..255] {     /* Memory Space */
     PORT { READ=1 OR WRITE=1 };
  }
  PROGRAM_MEMORY U32 ROM[0..255] {       /* Instruction ROM */
     PORT { READ=1 OR WRITE=1 };
  }
  PIN bit[1] RESET, U32 STATE_BUS;

  PIPELINE ppu_pipe = { FI; ID; EX; WB }; /* Pipeline Stage Definition for LISA */

  PIPELINE_REGISTER IN ppu_pipe  {
      bit[6] Opcode;
      S32 OperandA, OperandB, OperandC, OperandD, OperandE, OperandF;
      S32 ResultA, ResultB, ResultC, ResultD, ResultE;
  };
}
```

*Example 6.1:* Specification of the resource section of the ICORE architecture.

This information is sufficient to generate an equivalent basic architecture structure in a hardware description language like VHDL. Example 6.1 shows an excerpt of the resource section of the ICORE architecture.

Besides the general-purpose register file *R*, address registers (*AR*), the predicate registers (*PR*) and status registers (*SR*) are modeled. As the CORDIC algorithm is implemented using highly application-specific instructions which work simultaneously with a high number of variables, the general-purpose register file *R* is implemented with six read and six write ports. This is formulated using the PORT keyword (cf. chapter 3.2.1).

Since structure and behavior of the register file are fully specified, corresponding HDL code can be derived. Figure 6.4a shows a pictorial representation of the structure of the register file *R*, while figure 6.4b displays the resulting HDL code.

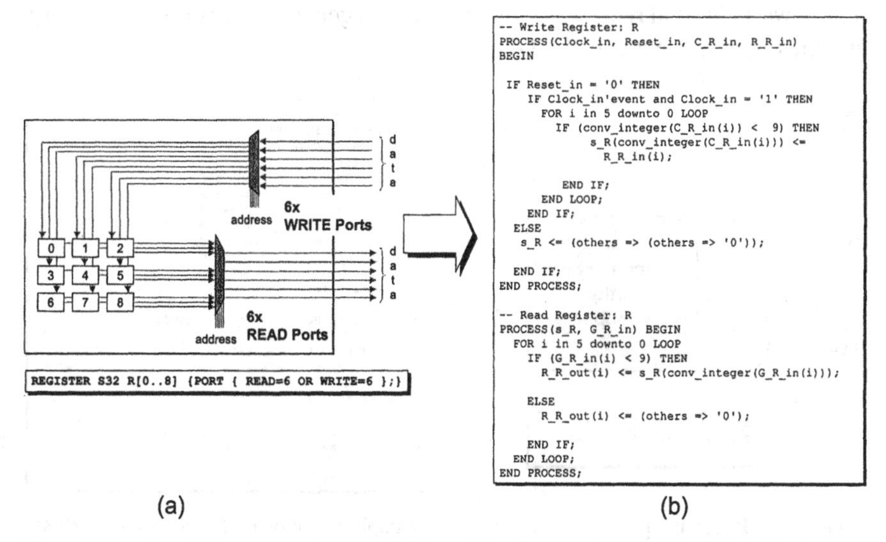


*Figure 6.4.* Pictorial representation and HDL code implementation of the register file *R*.

As already pointed out, the register file *R* is defined in the LISA description as an array with nine elements and six simultaneously allowed read and write accesses. These accesses are reproduced in the hardware, i.e. in the generated HDL code, as a loop that is traversed six times. All elements of the register file are set to zero in case the reset signal equals '*1*'. In normal operation mode, the register content (represented by the signal *s_R*) is written with the next rising clock edge. Furthermore, a data value is written to the respective register element based on the address on the *C_R_in* address bus. If this bus contains no valid register element number (i.e. the number is higher than eight) the value is not written. The read access to register file *R* is not sensitive to the clock edge. As soon as the address bus *G_R_in* changes its value, the correspondent register value is assigned to the data bus *R_R_out*. The naming conventions for the generated HDL code can be found in [147].

Except for the status register file *SR*, the further register files *AR* and *PR* allow one read or one write access per cycle (exclusiveness is indicated by the *XOR*, cf. chapter 3.2.2). As the register file *SR* contains no explicit information on access ports, it is assumed that all elements can be read and written simultaneously. Besides, registers to control the zero overhead loop (zolp) are declared in the model.

Moreover, a data and a program memory resource are declared – *RAM* and *ROM*, respectively – both 32 bits wide and with just one allowed read and write access per cycle. Since various memory types are known and are generally very technology dependent, however, cannot be further specified in the LISA model, wrappers are generated with the appropriate number of access ports. Figure 6.5 shows a pictorial representation of the entity generated for the data memory and the respective generated HDL code.

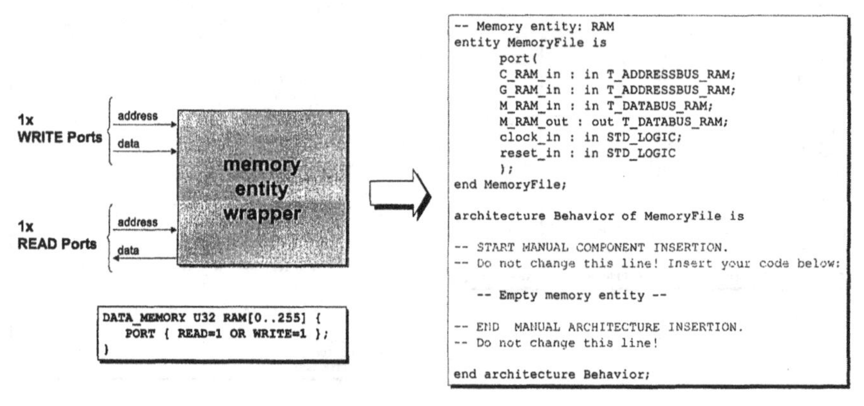


*Figure 6.5.* Pictorial representation and HDL code implementation of the data memory *RAM*.

The ports represent the specified number of data and address buses for read and write. Before synthesis, the wrappers need to be filled manually with code for the respective technology.

The resources labelled as PIN in example 6.1 are fed outside the top-level architecture entity and can be accessed via a test-bench – in the ICORE the *RESET* and the *STATE_BUS*.

For performance reasons, instruction execution is distributed over a four stage pipeline in the ICORE architecture. The stages are concerned with instruction fetch (FI), decode (ID), execute (EX), and write-back (WB). In addition, pipeline registers are assigned to the pipeline which are located in between the stages. The pipeline registers named *Operand* are located between the *ID* and *EX* stage and forward operand values from the decode to the execute stage. The operands are filled by the decoder in the *ID* stage. Moreover, the pipeline registers named *Result* are connecting the *EX* and *WB* stage.

In the hardware description language VHDL, a system is usually composed of a hierarchical collection of modules[2]. Modularization eases maintenance and reusability of the model. Each module has a set of ports which constitute its interface to the outside world. In VHDL, an entity is such a module which may be used as a component in a design, or which may be the top level module of the design.

From the information on the coarse structure of the architecture, a base model structure in VHDL can be derived. The respective base structure for the ICORE architecture is shown in figure 6.6.

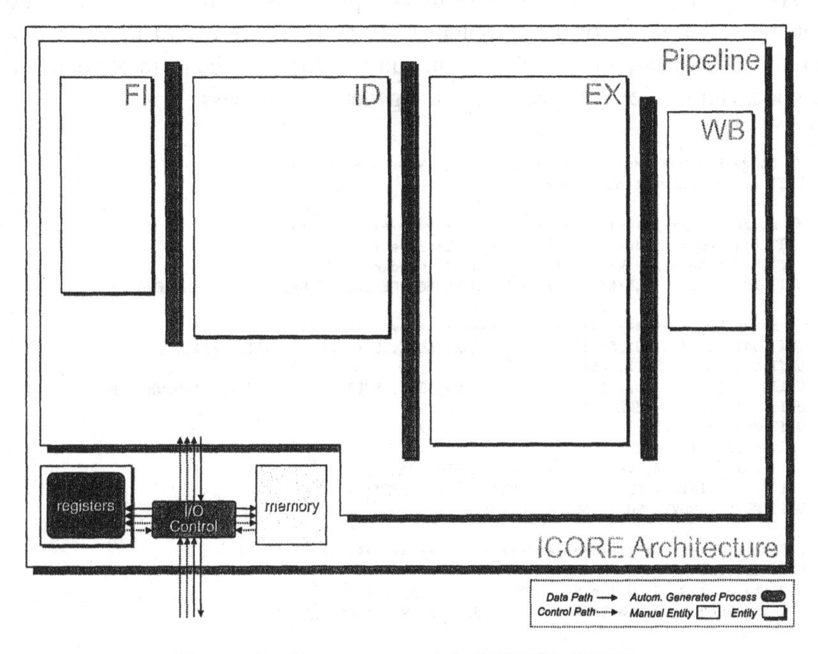


*Figure 6.6.* Base structure of the ICORE in VHDL.

The base structure of the generated VHDL model is made up by an architecture entity (top-level) which has three further entities embedded: register, memory, and pipeline. The register and memory entities contain further entities representing the declared registers and memories in the underlying LISA model (cf. figures 6.4 and 6.5 respectively). The pipeline stages and registers are embedded into the pipeline entity – for the ICORE architecture three pipeline register entities (between stages FI/ID, ID/EX, and EX/WB) and four stage entities are generated (for stages FI, ID, EX, and WB respectively). The

[2]LISA processor models are hierarchically structured as well, cf. chapter 3.1. However, hierarchy in VHDL is applied to structure, while in LISA it is applied to the description of the instruction-set.

I/O control allows access to those processor resources labelled as PIN in the LISA model.

Every generated entity is connected to a clock and a reset signal. The reset signal is treated by default in the whole architecture as an asynchronous reset. For the pipeline register entities, additional ports are generated to control their behavior. This concerns the issue of flushes and stalls by the pipeline controller. The pipeline control is discussed in section 2.4.

## 2.2 Grouping Operations to Functional Units

As indicated at the beginning of this chapter, it is not feasible to efficiently generate HDL code for the complete ASIP architecture from LISA processor models. The data-path, which is formulated in LISA in the form of behavioral C-code, is not taken into account in the generation process.

```
/* Stage FI: ********************************************/
UNIT Fetch_Instruction { Read_Instruction; }

/* Stage ID: ********************************************/
UNIT BranchDetectionUnit { CMP, CMPI, Absolute_Value; }
UNIT DAG { Read_PI, Write_PI, LAI, LAIRO, Cordic01; }
UNIT ZOLP { LPCNT,LPINI,RTS,B,BCC,BCS,BE,BG,BGE,BL,BLE,BNE,BPC,BPS,BSR,END; }

/* Stage EX: ********************************************/
UNIT Shifter { COR01_Shifter, COR2, SLA, SLAI, SRA, SRAI, SRAI1, SRA1; }
UNIT Bitmanip { RBIT, WBIT, WBITI; }
UNIT ALU { ADD, ADDI, AND, ANDI, NEG, OR, SUB, SUBI, XOR, ADDSUB0, ADDSUB1; }
UNIT Mult { MULS, MULU; }
UNIT Minmax { SAT; }
UNIT Addsub { COR01_AddSub; }
UNIT Mem { RD, RA, WR, WA, RPI, WPI, COR01_Mem; }
UNIT IIC { INI2C, INA, OUTI2C, OUTA, INPI, OUTPI; }
UNIT MOVE { MOVI, MOV, ABS, COR01_Move; }

/* Stage WB: ********************************************/
UNIT WriteBack
{ Write_Back_COR2, Write_Back_COR01, Write_Back_A,
  Write_Back_B, Write_Back_C, Write_Back_D, Write_Back_E; }
```

*Example 6.2:* Declaration of functional units in the ICORE architecture.

The HDL code of the data-path has to be added manually after the generation process. In analogy to the treatment of memories in the model (cf. section 2.1), wrappers are generated by the LISA environment representing functional units which can be filled with the respective HDL code. Moreover, the interconnections from the functional units to processor resources (e.g. pipeline-registers and memories) are generated automatically from information provided by the underlying LISA processor model. As several hardware operations can share one functional unit, the LISA language provides a mechanism to group operations to units. The UNIT keyword is embedded in the resource section and allows the specification of functional units in conjunction with LISA operations assigned to the respective unit. Example 6.2 shows an excerpt of the LISA model for the ICORE architecture with the declaration of units.

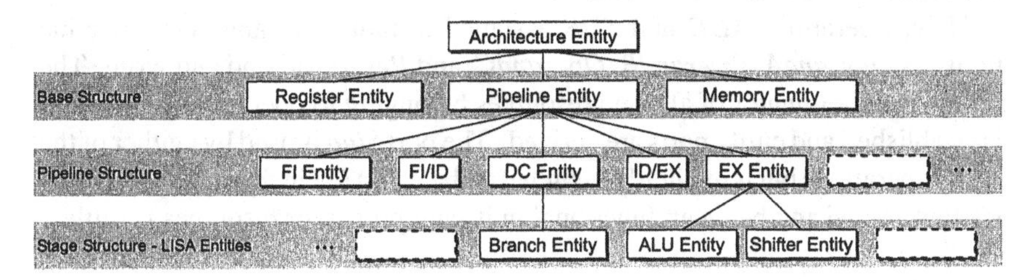


*Figure 6.7.* Entity hierarchy of the generated HDL model.

The location of functional units within the model is implicitly fixed by the assignment of LISA operations to pipeline stages in the model (cf. chapter 3.2.5 – *timing model*). The HDL representation of these functional units are again VHDL entities which are included into the respective stage entities. Figure 6.7 shows an excerpt of the resulting entity hierarchy of the generated HDL model.

Here, the hierarchies for the unit *Branch* assigned to stage *ID* and units *ALU* and *Shifter* assigned to stage EX are exemplarily shown. For simplification, the entities for the remaining stages (i.e. IF and WB) and units are not shown.

LISA Model

```
UNIT ALU { Add, Sub; USES READ PIPELINE_REGISTER(pipe, DC/EX).Sign;}

OPERATION Add IN ppu_pipe.EX {
  ...
  BEHAVIOR(USES READ   PIPELINE_REGISTER(pipe, DC/EX).OperandA,
                       PIPELINE_REGISTER(pipe, DC/EX).OperandB;
                 WRITE PIPELINE_REGISTER(pipe, EX/WB).Result;)
  {...}
}

OPERATION Sub IN ppu_pipe.EX {
  ...
  BEHAVIOR(USES READ   PIPELINE_REGISTER(pipe, DC/EX).OperandA,
                       PIPELINE_REGISTER(pipe, DC/EX).OperandC;
                 WRITE PIPELINE_REGISTER(pipe, EX/WB).Result;)
  {...}
}
```

VHDL

```
Entity ALU IS
   PORT ( OperandA  : IN  TypeOfOperandA;
          OperandB  : IN  TypeOfOperandB;
          OperandC  : IN  TypeOfOperandC;
          Sign      : IN  TypeOfSign;
          Result    : OUT TypeOfResult;
        );
END ALU;
```

*Figure 6.8.* Generation of entity ports from information on used resources in LISA.

The ports of the entities are derived from information on used resources in the behavioral code of the operations in the LISA model (cf. chapter 3.2.2 – *resource model*). This is formalized by the keyword USES in conjunction with the name of the used resources and the information if the resource are read, written or both. Figure 6.8 shows sample LISA code and the resulting port declaration of the entity in VHDL.

Here, operations *ADD* and *SUB* assigned to unit *ALU* announce pipeline registers *OperandA*, *OperandB*, *OperandC*, and *Result* for read and write. The common denominator of all used resources belonging to operations of one unit is established and entity ports are derived. The port *Sign* is used by neither of the LISA operations but is explicitly assigned to the unit *ALU* in the unit declaration. By this, the wiring between functional units and processor resources is settled.

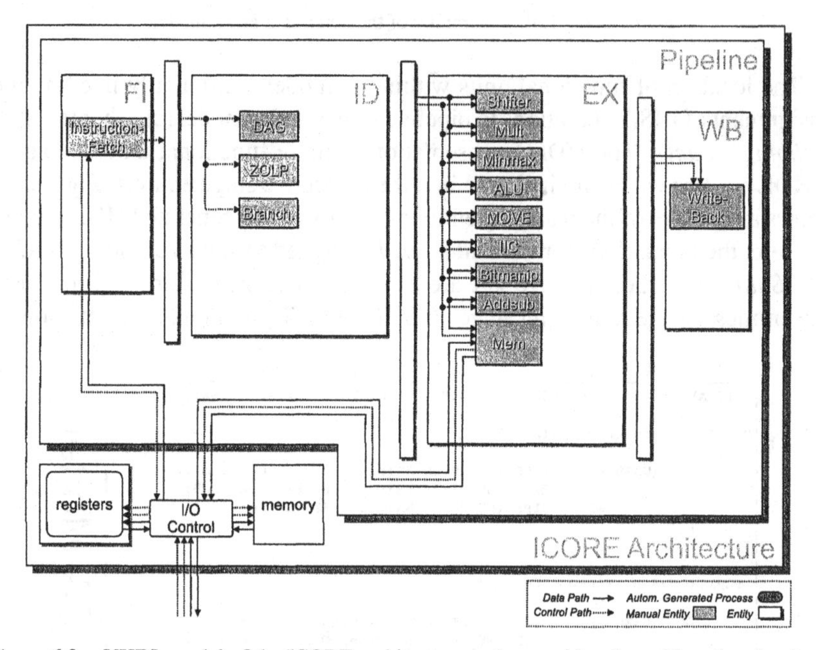


*Figure 6.9.* VHDL model of the ICORE architecture under consideration of functional units.

Figure 6.9 shows the structure of the VHDL model after consideration of units and their wiring to processor resources.

## 2.3 Decoder Generation

The decoder in the HDL model can be derived from information on the correlation between the coding of instructions (cf. chapter 3.2.4 – *instruction-set model*) and the activation of operations (cf. chapter 3.2.5 – *timing model*) in the underlying LISA model.

Depending on the structuring of the LISA model (i.e. the mode of operation activation), decoder processes are produced in pipeline stages. Figure 6.10

shows the coarse operation activation chain in the ICORE architecture and the extraction of decoder processes.

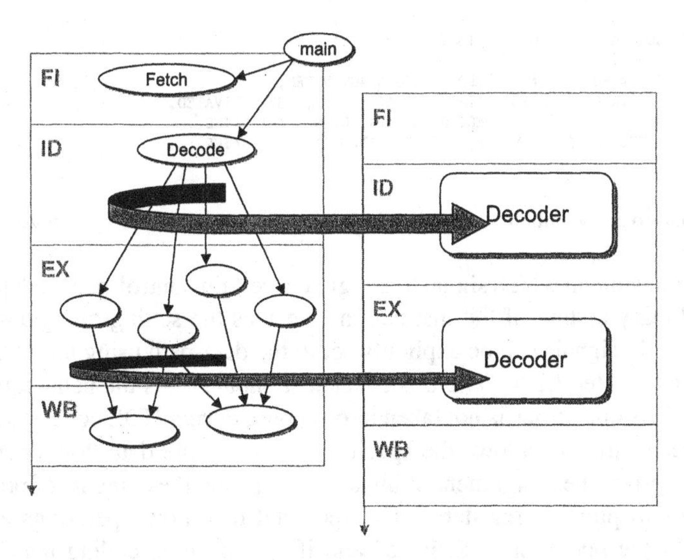


*Figure 6.10.* Location of decoder processes in the operation activation tree.

Here, operation *main* activates operations *Fetch* and *Decode* every cycle. As this process is always active, no decoder is required. Starting in stage *ID*, operation activation is dependent on the coding of instructions. Thus, decoders are inserted into the generated HDL model. The signals set by the decoder can be divided into those accounting for the control and those for the data-path. The control signals are a straight forward derivation of the operation activation tree while the data signals are explicitly assigned within LISA operations.

In LISA, operations are scheduled for execution using the activation mechanism. Activated operations are scheduled for execution in the pipeline stage they are assigned to. In chapter 3.2.5 it was argued, that the activation mechanism can be compared with setting signals in hardware. Indeed, for every activation sequence in the model, a physical control-path is created in the generated HDL model. This control-path concept in the HDL is the equivalent to the LISA activation mechanism. The translation procedure is characterized by two tasks that need to be carried out by the HDL code generator:

- creation of control-path triggering functional units, and

- derivation of processes setting the control-path to its actual value according to the currently executed instruction.

The second task is carried out by the decoder. Example 6.3 shows the valid control signals generated by the LISA environment to trigger the operations activated from operation *ALU_operations*.

```
TYPE T_ALU_Operations_Opcode IS (

IS_NOT_ACTIVATED, ADD_ACTIVATED, ADDSUB0_ACTIVATED,
ADDSUB1_ACTIVATED, AND_ACTIVATED, CMP_ACTIVATED, MOV_ACTIVATED,
MULS_ACTIVATED, MULU_ACTIVATED, OR_ACTIVATED, SLA_ACTIVATED,
SRA_ACTIVATED, SRA1_ACTIVATED, SUB_ACTIVATED, XOR_ACTIVATED);
```

*Example 6.3:* Definition of control-path signal set by operation *ALU_operations*.

The decision on which signal to assign to a certain control-path is dependent on the binary coding of the instruction. Besides the setting of signals to the control-path, signals can be explicitly set to the data-path using the SET statement (cf. chapter 3.2.6 – *micro-architectural model*). As the behavioral code within LISA operations is not taken into account during HDL code generation, this mechanism still allows the specification of simple data-flow operations. This comprises the assignment of binary codings and the contents of processor resources to pipeline registers. The specified data-flow operations are performed if the operation is activated and if the specified coding matches the current instruction.

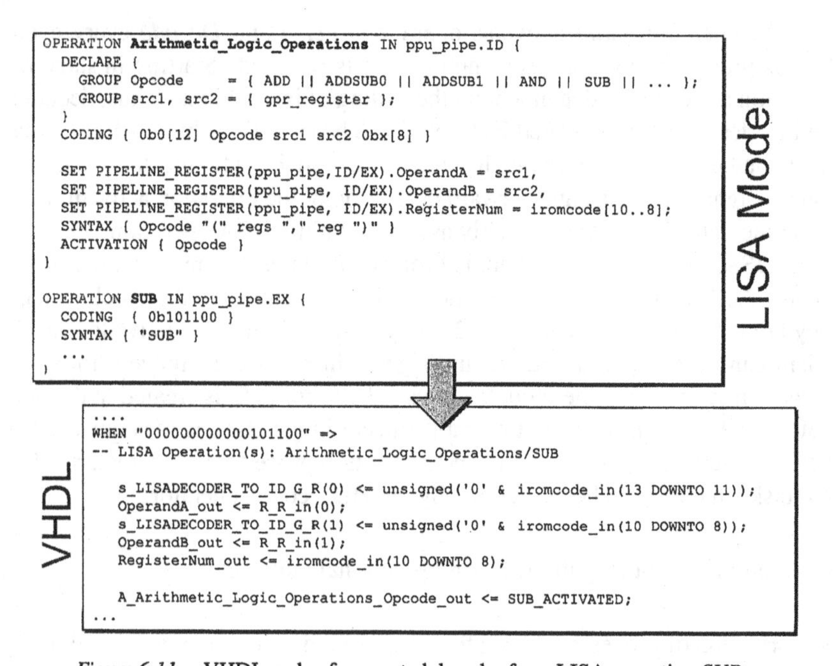


*Figure 6.11.* VHDL code of generated decoder from LISA operation *SUB*.

Figure 6.11 shows sample LISA code taken from the ICORE architecture modeling the subtraction instruction and the generated VHDL code. In the LISA model, the instruction is split up into two operations – *Arithmetic_Logic_Operations* and *SUB*. Operation *Arithmetic_Logic_Operations* is selected by the instruction decoder in case the instruction word has twelve zeros in the most significant bits (MSB). Moreover, the register values of *src1* and *src2* are written into the respective pipeline registers *OperandA* and *OperandB* using the SET statement. Besides, pipeline register *RegisterNum* is loaded with an excerpt of the instruction word. Depending on the coding of the *Opcode* field, the respective operation is activated. In the example, operation *SUB* is activated if the *Opcode* field equals *101100.*

The generated VHDL code representing a part of the decoder is shown in the lower part of figure 6.11. Here, the respective control and data signals are set. The entity port *A_Arithmetic_Logic_Operation_ACTIVATED* is set to a value indicating that the *SUB* instruction has been decoded. Additionally, the register values are assigned to the respective pipeline registers. This procedure is carried out in two steps: firstly, the addresses of the registers are written to the register address bus. As the signals and ports change their values at the end of processes by definition, the register data bus does not contain the correct register values in the first run. The decoder process is sensitive to the register data bus and thus the decoder is worked out a second time. This time, the register data bus contains the correct values for *OperandA* and *OperandB*. This behavior is shown in the timing chart in figure 6.12.

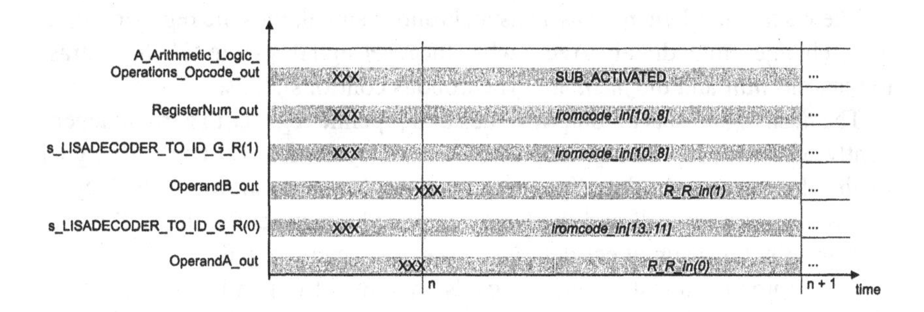


*Figure 6.12.* Timing chart for decoded SUB instruction.

As can be seen, the ports *OperandA_out* and *OperandB_out* do not change their value with the first decoder run at time *n*.

## 2.4 Pipeline Control

For resolution of hazards, the LISA environment provides a set of predefined pipeline control functions (cf. chapter 3.2.5 – *timing model*). These comprise pipeline register *stalls*, *flushes*, and *inserts* and can be applied to single pipeline registers or to all pipeline registers simultaneously.

In the generated HDL model, all pipeline registers are equipped with four input control ports, in addition to the ports for the control and data-path set by the decoders. In normal operation mode, the *out* ports of the pipeline registers take over the value of the *in* port with the next rising clock edge. Figure 6.13 shows a sample timing chart of a pipeline register element.

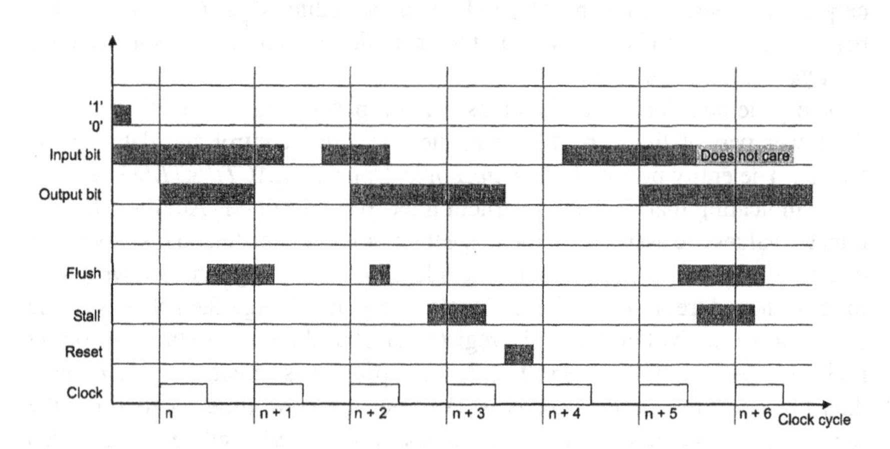


*Figure 6.13.* Timing chart for sample pipeline register element.

The *reset* is implemented as an asynchronous signal, thus the register output value changes immediately to zero when the reset signal is set to '1'. In contrast to that, the *stall* and the *flush* are synchronous control signals.

The *flush* signal sets all output values of a pipeline register to zero, independently of the input value as indicated at clock cycle $n + 1$. The *stall* signal retains the old output values, ignoring the input values, as shown in clock cycle $n + 3$. In case the stall and the flush are issued concurrently, the stall signal precedes the flush signal per definition (see clock cycle $n + 6$).

As the pipeline register control signals are issued from within the activation section in the LISA description they are set by the generated decoder like any other control signal in hardware. These special signals are directly connected to the pipeline controller.

Figure 6.14 shows the structure of the VHDL model of the ICORE under consideration of the pipeline controller and multiplexers. The pipeline controller transmits the respective control signals to the addressed pipeline registers. In fact, the controller combines all control signals issued from decoders inside the pipeline and passes them on to the target pipeline registers under consideration of priorities.

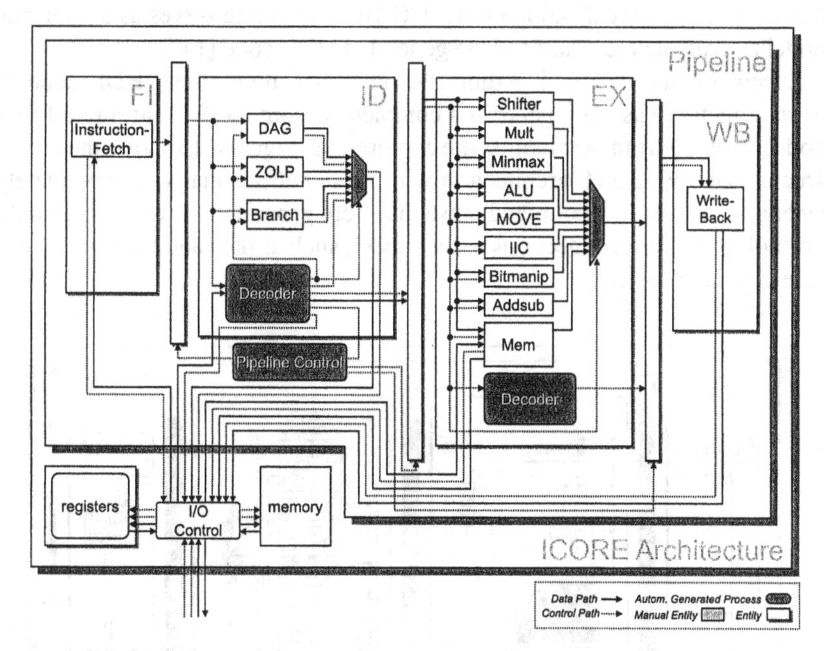


*Figure 6.14.* VHDL model of the ICORE under consideration of pipeline control.

Another important issue for the HDL code generation is the automatic insertion of multiplexers into the model. In the ICORE architecture, all functional units in the *EX* stage write to at least one of the pipeline registers elements carrying the result. The control signals derived from the activation section and set by the generated decoders steer the multiplexers and thus determine which functional unit is allowed to write to the respective register.

## 3. Case Study

The ICORE architecture, which serves as a case study for the architecture implementation using LISA, was originally developed at the Institute for Integrated Signal Processing Systems (ISS) at Aachen University of Technology in cooperation with Infineon Technologies. Today, it is available as a product being part of a Set-Top Box [148].

In the first realization of the architecture, exploration was carried out using the LISA methodology and tools (cf. chapter 5). However, as the path to implementation was not available at this time, i.e. the HDL code could not be generated automatically from the LISA model resulting from the exploration, the architecture was implemented manually in semi-custom design. Thereby, a large amount of effort was spent to optimize the architecture towards extremely low power consumption while keeping up the clock frequency at 125 MHz. This manual, optimized realization of the ICORE architecture serves as a reference model to evaluate the quality of the generated HDL code [147].

Except for the data-path within the functional units, the VHDL code of the architecture has been generated completely from the RTL-accurate LISA model resulting from the architecture exploration. Figure 6.15 shows the coarse structure of the ICORE architecture with those entities that were completely generated in dark grey color and those that were generated as wrappers in light grey color. All interconnections in the model, such as data and control signals, were generated as well.

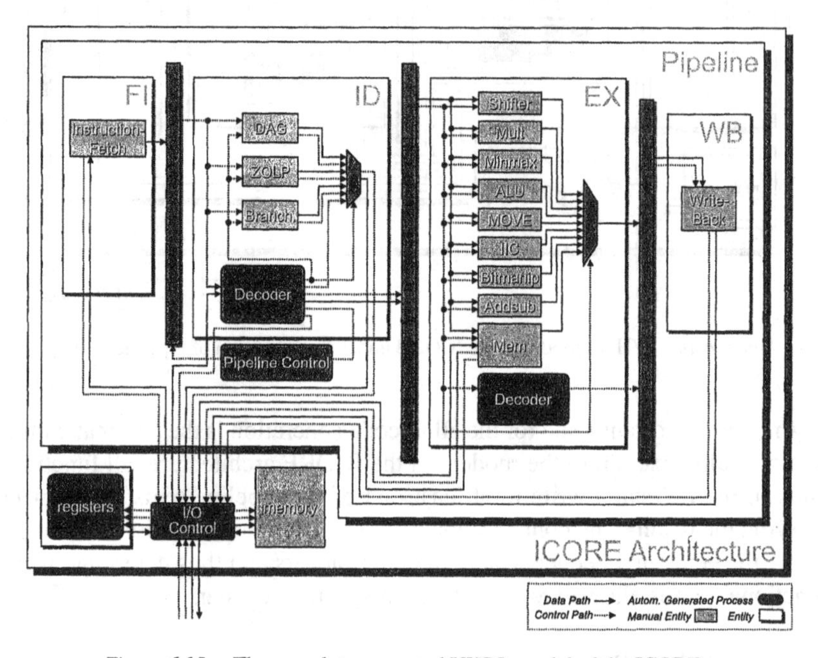


*Figure 6.15.* The complete generated VHDL model of the ICORE.

## 3.1 Model Development Time

The LISA model of the ICORE as well the original handwritten HDL model of the ICORE architecture have been developed by one designer. The initial manual realization of the HDL model (not considering the time needed for architecture exploration) took approximately three months. As pointed out above, the LISA model was already developed in this first realization of the ICORE for architecture exploration and verification purposes.

It took the designer approximately one month to learn the LISA language and to create a cycle-based LISA model. After completion of the HDL generator, it took another two days to refine the LISA model to the level of RTL micro-architectural accuracy. The handwritten functional units (data-path), that were added manually to the generated HDL model, could be completed in less than a week.

As this comparison indicates, the time-consuming work in realizing the HDL model was to create structure, controller, and decoder of the architecture. In addition, a major decrease of total architecture design time can be seen, as the LISA model results from the design exploration phase.

## 3.2 Comparison of Structure and Size

After running the HDL code generator on the basis of the LISA model, the model sizes of generated and handwritten model were compared. The size was quantified by counting the number of lines of code in both models. The reason for not making the comparison on the basis of the number of characters in each file (i.e. the binary size of the file) is that the names of signals and ports in the generated file are much longer than the handwritten equivalents. This is because the developer needs a higher level of transparency in an automatically generated code. Otherwise, manual optimization of the generated code would be extremely tedious.

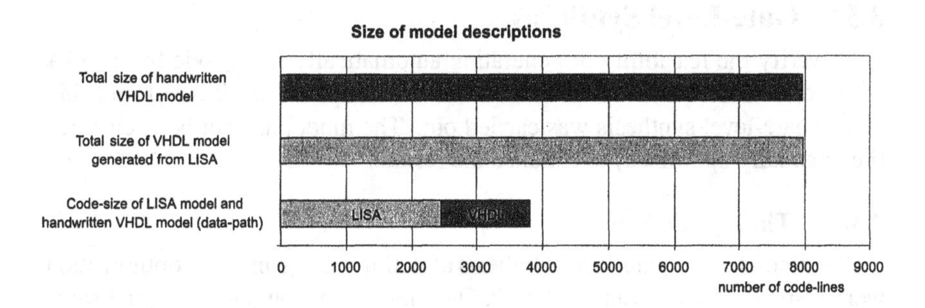


*Figure 6.16.* Code Size of the LISA description and the VHDL models.

Figure 6.16 displays the sizes of different models of the ICORE architecture. The LISA model of the ICORE comprises 2457 lines of code. The number of lines required to complete the HDL code generated from this LISA model (i.e. the code for the data-path) turns out to 1356 lines of VHDL code. The total code size of the handwritten and the automatically generated VHDL model are almost identical with 7950 and 7953 lines respectively. As a matter of fact, these numbers are of no significance for the quality of the model in terms of area and power consumption.

Setting the amount of VHDL code generated automatically from the underlying LISA model in relation to the amount of VHDL code added to complete the functional units, it is interesting to notice that the architecture has been generated by 82,95% (see figure 6.17).

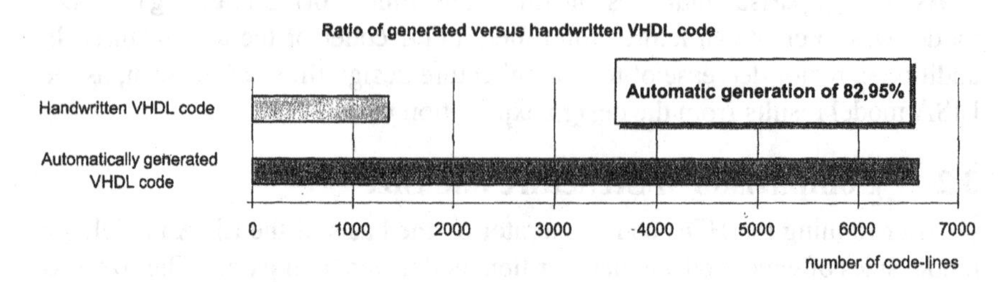


*Figure 6.17.* Percentage of automatically generated code.

This proves the hypothesis that most effort has to be spent on the realization of the coarse structure and control of the architecture, which is tedious to write manually although it does not play a major role for the performance. On the other hand, the performance critical parts of the architecture, which need the attention of the designer, only account for less than 20%.

## 3.3 Gate-Level Synthesis

To verify the feasibility of generating automatically HDL code from LISA architecture descriptions in terms of power-consumption, clock speed, and chip area, a gate-level synthesis was carried out. The model has not been changed (i.e. manually optimized) to enhance the results.

### 3.3.1 Timing and Size

The results of the gate-level synthesis affecting timing and area optimization were compared to the handwritten ICORE model, which comprises the same architectural features. Moreover, the same synthesis scripts and technology were used for both models.

It shall be emphasized that the performance values are nearly the same for both models. Furthermore, it is interesting to notice that the same critical paths were found in both, the handwritten and the generated model. The critical paths occur exclusively in the data-path, which confirms the presumption that the data-path is the most critical part of the architecture and should thus not be generated automatically from a more abstract processor model.

**Critical Path.** The synthesis has been performed with a clock of 8ns – this equals a frequency of 125MHz. The critical path, starting from the pipeline register between stages *ID/EX* and going through the shifter unit and multiplexer to the next pipeline register between stages *EX/WB*, violates this timing constraint by 0.36ns. This matches the handwritten ICORE model, which has been improved from this point of state manually at gate-level.

The longest combinatorial path of the *ID* stage runs through the decoder and the DAG entity and counts 3.7ns. Therefore, the generated decoder does not affect the critical path in any way.

**Chip Area.** The synthesized area has been a minor criterion, due to the fact that the constraints for the handwritten ICORE model are not area sensitive. The total area of the generated ICORE model is 59009 gates. The combinational area takes 57% of the total area. The handwritten ICORE model takes a total area of 58473 gates.

The most complex part of the generated ICORE is the decoder. The area of the automatically generated decoder in the *ID* stage is 4693 gates, whereas the area of the handwritten equivalent is 5500 gates.

### 3.3.2 Power Consumption

Figure 6.18 shows the comparison of power consumption of the handwritten versus the generated ICORE realization.

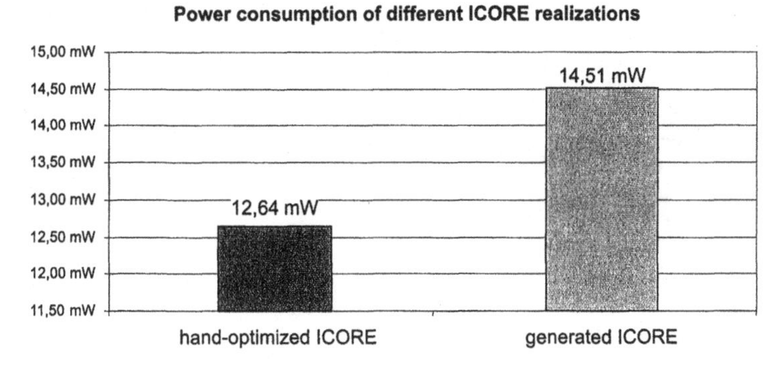


*Figure 6.18.* Power consumption of different ICORE realizations.

The handwritten model consumes 12,64mW whereas the implementation generated from a LISA model consumes 14,51mW. The reason for the slightly worse numbers in power consumption of the generated model versus the handwritten is due to the early version of the LISA HDL generator which in its current state allows access to all registers and memories within the model via the test-interface. Without this unnecessary overhead, the same results as for the hand-optimized model are achievable.

## 4. Concluding Remarks

In this chapter, the second quadrant of the LISA processor design platform was introduced – architecture implementation. It was shown that it is feasible to generate synthesizable HDL code from an RTL micro-architectural LISA processor description for most parts of the architecture. This comprises the coarse structure including register banks, pipelines, pipeline controllers as well as instruction decoders setting control and data signals to steer instruction execution. For the functional units within the architecture (i.e. the data-path), wrapper entities are generated which are filled manually with optimized HDL code for the target architecture.

In a case study, the ICORE architecture was presented, which was realized using the LISA methodology and tools. It was shown that area and power consumption of the automatically generated architecture can compete with a separate version of the same architecture that was optimized manually. This proves the proposed concept of generating those parts of the architecture automatically that are lengthy and tedious to specify, but only have negligible influence on the performance of the architecture, while leaving those parts that are critical untouched.

# Chapter 7

# SOFTWARE TOOLS FOR APPLICATION DESIGN

Once the architecture design is finished, a set of production quality software development tools is required to comfortably program the target architecture [149, 150]. However, as the demands of the application designer on the software development tools partially contradict those of the hardware designer (cf. chapter 4.2), a new set of tools is needed.

For the application software design phase, production quality software development tools comprising assembler, linker, simulator, and debugger can be generated automatically from LISA processor models. Here, *production quality* refers to the functional ranges, speed, and graphical debugging capabilities that are comparable to what commercially available software development tools offer. The tools are derived from the same models that were used during the architecture exploration and implementation phase. In the majority of cases, the instruction-based model is chosen for this purpose, as it provides the required information on the instruction-set architecture concerning binary coding and assembly syntax while keeping up simulation speed at a high level due to the degree of abstraction (cf. chapter 5.1). Indeed, instruction-based models provide the bit-accurate behavior of the processor. With regard to the timing behavior, any instruction is executed within one control step during simulation. However, in most cases, instruction-based models mirror the cycle-count required to execute a certain instruction. For this reason they are also often referred to as cycle-count accurate models[1].

Figure 7.1 shows the generation process and the tools required to benchmark the application on the target architecture during the application software design phase. The generated assembler *lasm* processes textual assembly source-code

[1]Note: in contrast to that, cycle-based models also describe pipelines and the instruction distribution over several pipeline stages, cf. chapter 5.1.

of the application under test and transforms it into linkable object code for the target architecture. The transformation is characterized by the instruction-set information on binary coding and syntax defined in the LISA processor description. Following that, the linker *llnk* assigns the relocatable output of the assembler to the processor's address space. This step is driven by a command file which contains the target architectures' memory configuration. The generated executable can be loaded into the software simulator which employs two simulation techniques – interpretive and compiled. The choice of the simulation technique applied is based on both architectural and application characteristics (cf. section 2.1). The simulator core can be coupled to the graphical debugger frontend *ldb* which visualizes the application run and controls the simulator. To increase the debug ability of the application code, the binary executable is disassembled by the disassembler *ldasm* and displayed in the graphical debugger frontend.

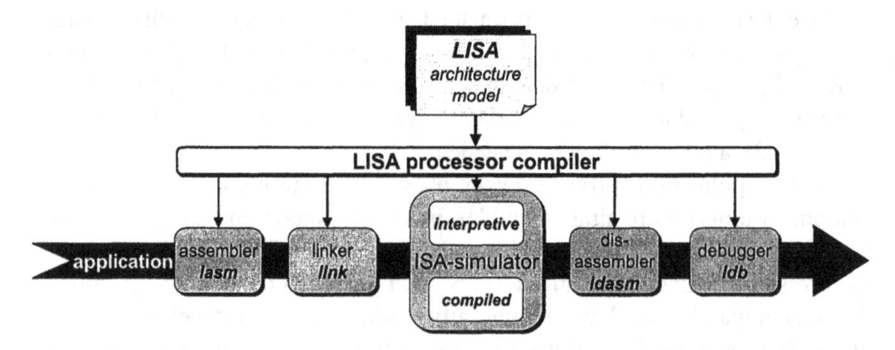


*Figure 7.1.* Software development tools for application design.

This chapter introduces the third quadrant of the LISA processor design platform – software development tools for application design. Thereby, emphasis is laid on the specialties and requirements of the respective software development tools with a particular focus on different simulation techniques. Starting with the code generation tools, assembler and linker generated from LISA processor models are briefly introduced. Subsequently, different simulation techniques supported by the LISA processor design platform are presented. This concerns on the one hand the most commonly used interpretive simulation technique, and on the other hand the compiled simulation technique of two different flavors – dynamically and statically scheduled. Moreover, the trade-off between these simulation techniques with respect to simulation speed, simulation compilation time, and flexibility is discussed. The chapter closes with four case studies showing benchmarking results for real-world processor architectures which were modeled using LISA and for which tools were successfully generated.

# 1. Code Generation Tools

The software development tools generated from LISA architecture descriptions provide a comprehensive development environment starting with application development on the assembly level. However, it is obvious that using target-specific program descriptions has several shortcomings:

- programming assembly is a lengthy, tedious, and error-prone process especially in the context of architectures with explicit instruction-level parallelism (e.g. Texas Instruments TMS320C6x [15], Motorola StarCore [151], and Analog Devices TigerSHARC [152]) that require techniques like software pipelining,
- the verification effort of assembly programs is immense and contradicts the goal of enabling fast exploration of the application and target architecture, and
- assembly sources are hard to reuse when porting from one to the next architecture and mostly require a complete rewriting of the application from scratch.

The common solution to these problems is the usage of a high-level language (HLL) compiler that enables the translation of an abstract and architecture independent algorithmic description (e.g. a C program) into the architecture dependent assembly code. Obviously, the compiler must consider the architecture and therefore be modified each time the architecture is changed during exploration [153].

Besides, one of the key requirements on compilers is to generate assembly code of acceptable quality. As today's architectures are mostly memory dominated, a new metric to minimize chip area – *code density* – comes into focus. It was shown in [154], that even commercially available, hand-optimized C-compilers for DSP architectures are only capable of delivering code that exceeds the size of hand-optimized assembly code by an order of magnitude. These numbers were gathered in 1996, however, it was pointed out in [155] that the situation has not improved.

Work on retargetable compilers parameterized from LISA architecture descriptions is at an early stage of research [156] and not covered within this book (cf. chapter 9.3). At the current stage of research, application design starts at the assembly level.

## 1.1 Assembler Generation

Assemblers play a major role in application design. On the one hand, HLL compilers require assemblers to process the compiler output. On the other hand, time critical parts of the application such as inner loops are still realized at the assembly level, even at the presence of a HLL compiler. Besides, under certain circumstances, compilers are not suitable at all, especially if the application to be executed is very small and the instruction-set highly non-orthogonal [157]. This is frequently the case in ASIP architectures that trade programmability and compiler friendliness through orthogonality against power efficiency and simplicity.

Here, an assembler offers a significant programming easement by a first abstraction of the hardware. It translates text-based instructions into object code for the respective programmable architecture. In addition to that, symbolic names for opcodes (*mnemonic*), memory-contents (*variable*), and branch addresses (*label*) simplify programming. Besides the processor-specific instruction-set, the assembler provides a set of pseudo-instructions to control the assembling process (*directives*). This concerns data initialization, reasonable separation of the program into sections (*section*), handling of symbolic identifiers for numeric values and branch addresses.

The assembler generated from LISA processor models has to perform two major tasks. Firstly, the parsing and processing of all instructions of the assembly application according to the description of the instruction-set in the underlying LISA description (cf. chapter 3.2.4). This concerns the transformation of the assembly syntax into the respective binary coding and is thus a highly architecture dependent step. Secondly, directives, labels, and constant symbols require handling in the assembly application. These are recurring tasks that are independent of the target architecture.

For the above mentioned reasons, the target-specific assembler is composed of the following two components:

- a *dynamic* part, which is generated from the LISA processor model comprising the instruction parser and a translation unit that compiles the assembly code into binary instruction words, and
- a *static* part, which is independent of the target architecture and is responsible for handling directives, labels, and symbols. Apart from that, the assembly code is assigned to different sections according to the specification in the assembly file.

These two parts are linked together into a stand-alone application representing the target-specific assembler. In the following, the work-flow of the generated two-pass assembler is briefly illustrated.

**Work-flow of the retargetable assembler** The work-flow of the assembler generated from LISA processor models can be divided into two major passes as shown in figure 7.2.

In the first pass, the syntactical correctness of the assembly program is checked. Moreover, the word-sizes of the processor instructions are determined

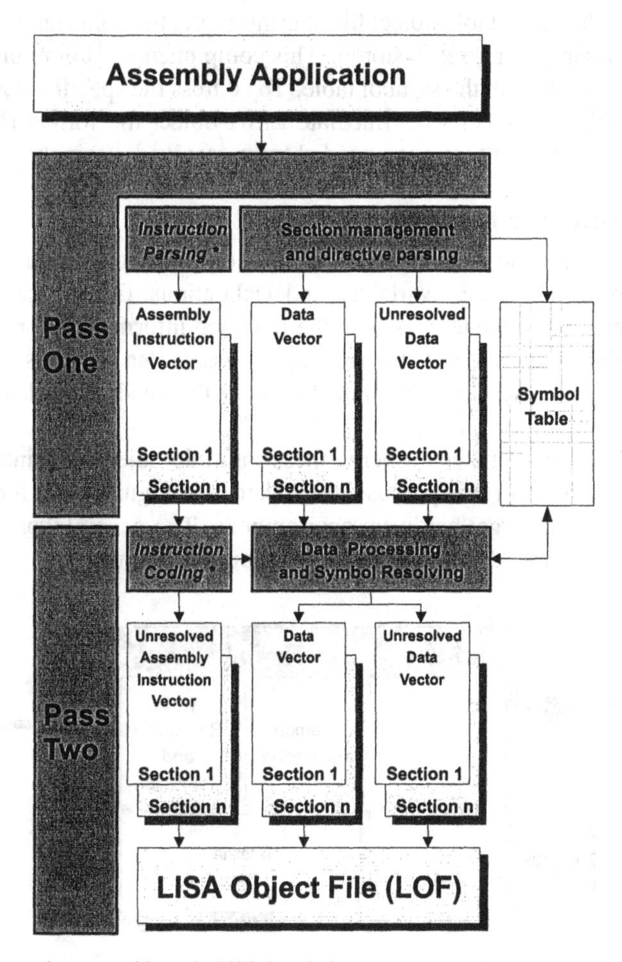


*Figure 7.2.* Work-flow of the two-pass assembler generated from LISA processor models.

for the address calculation, a symbol-table is created, and data is initialized which is defined in the assembly program.

In the second pass, the instructions are transformed into object code. At this point in time, all absolute symbols are available in the symbol-table and can be resolved for assembling the instructions which make use of symbolic identifiers. As no binding addresses are available during the assembling process, all addresses depend on the memory location which the program is to be linked to. Consequently, labels cannot be resolved when assembling – the appropriate object code is left blank and will be filled in the linking pass. These relocation entries are stored separately within the intermediate output of the assembler – the relocatable object file. Furthermore, the collected information of the two assembler passes is stored. This comprehends section information, binary object code, and the symbol table. To address the specific requirements of retargetability, a special intermediate LISA object file format (LOF) was designed to store the information needed for unresolved assembly [158].

## 1.2 Linker Generation

Large programs consist of multiple modules, of which each module is a unit of logically grouped functions, variables, and declarations. Each of these modules can be assembled separately, as only the interface information is required for the assembling process. For this reason, pre-assembled modules, as e.g. for I/O, mathematics, etc. can be shipped without the need to disclose also the source-code.

It is the task of a linker to merge these multiple object files into a single executable object. For this purpose, the linker is configured with the memory information of the target hardware environment. The general work-flow of a linker can be separated into four passes as depicted in figure 7.3.

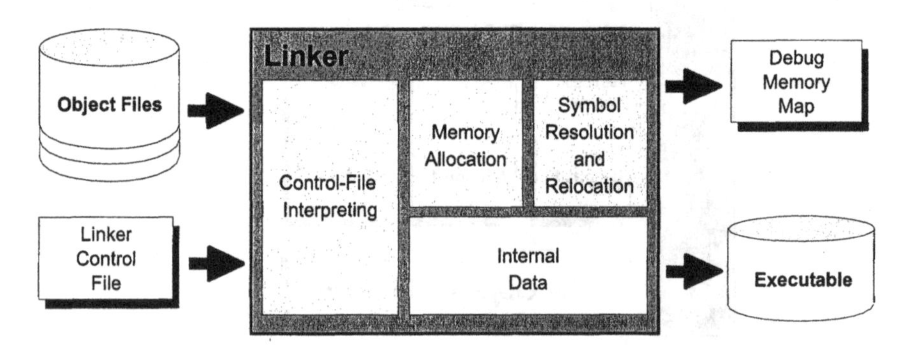


*Figure 7.3.* General work-flow of linkers.

In the first pass, an internal representation of the memory model is built up and the input object files are loaded. After that, in the second pass, the respective memory is allocated for the different sections. Following the memory allocation process, the address of each section is fixed. Thus, in the third pass, the relative symbols (*labels*) are resolved and the relocation entries (cf. section 1.1) are filled. The fourth pass is characterized by the creation of the executable output object file.

**Work-flow of the retargetable linker** There are high requirements on the flexibility of the linker. This is due to the fact that LISA was designed for the description of both DSP and $\mu$C architectures, which can be embedded into systems with arbitrary memory configurations.

Therefore, the retargetable linker provides the facilities to adjust an executable to the target memory environment. However, the underlying LISA model only provides information about internal memories and/or external memory configurations that are fixed at compile-time of the processor model. As this mirrors an unacceptable limitation, the linker includes additional information about the target memory configuration at the time the linking takes place. For this purpose, a linker *command-file* is provided which controls the linking process and assigns program and data sections to dedicated memory locations. Figure 7.4 illustrates a sample section composition within the executable output file.

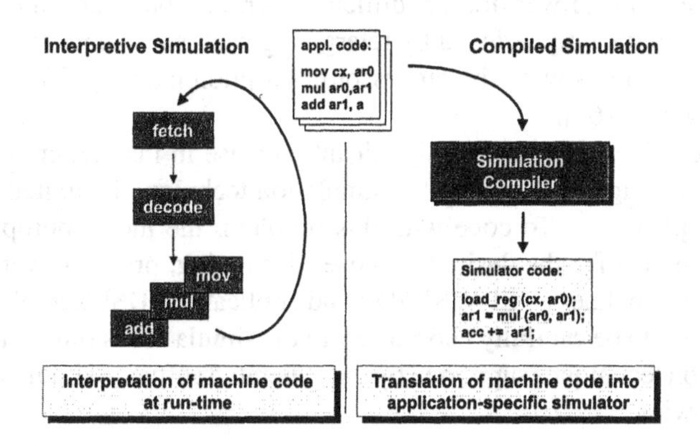


*Figure 7.4.* Section composition from LISA linker command file.

The retargetable linker creates an executable COFF [159] object module from multiple input object files in the LOF format (cf. section 1.1). As specified in the command-file, the linker combines the sections from the input-file (input-sections) to output-sections which are to be written to the executable output file. These output-sections are then linked to a binding address in a configured memory. The linker watches memory conflicts like overlapping sections and resource availability. Moreover, it supports named memory, paging, and automatic placement of sections into memory. The command-file syntax is derived from the Texas Instruments TMS320C6x tools [160].

## 2. Simulation

Due to the large variety of architectures and the facility to develop models on different levels of abstraction in the domain of time and architecture (cf. chapter 3), several simulation techniques ranging from the most flexible *interpretive* simulation to more application- and architecture-specific *compiled* simulation techniques are employed in the generated software simulator.

Compiled simulators for programmable architectures have been proven to outperform the commonly used interpretive simulators by roughly two orders in magnitude in speed without any loss in accuracy [161]. The principle of compiled simulation takes advantage of a priori knowledge and moves frequent operations from simulation run-time to compile-time in order to provide the highest possible simulation speed. Between fully compiled and fully interpretive simulation different levels of compiled simulation can be distinguished, ranging from the mere compile-time instruction decoding up to compile-time scheduling for pipelined processor models (*static scheduling*). In the past, such simulators have been realized manually for specific processor architectures [162]. However, compiled simulators for programmable architectures are complex pieces of software that are difficult to write, debug, and validate. These efforts are significantly reduced by retargeting such simulators from machine descriptions – in this work the processor description language LISA was used for this purpose [163].

Compiled simulators offer a significant increase in instruction throughput, however, the usage of the compiled simulation technique is limited to certain areas of application. To cope with this problem, the most appropriate simulation technique for the desired purpose (debugging, profiling, verification), architecture (pipelined, VLIW, SIMD), and application (DSP kernel, operating system) needs to be carefully chosen before the simulation is run. An overview of simulation techniques supported by the generated software simulator is given in the following.

The *interpretive* simulation technique is employed in most commercially available instruction-set simulators (e.g. [164] of Analog Devices, [165] of Texas Instruments and [166] of Advanced Risc Machines). In general, interpretive simulators run significantly slower than compiled simulators. This is due to the fact that interpretive simulators operate similarly like the real hardware does, i.e. prior to their execution instructions they are fetched from memory and decoded (see figure 7.5). However, especially for DSP applications containing code with a high locality this technique is suboptimal, as the time-consuming process of decoding is performed several times for the same instruction.

The advantage of this simulation technique is its flexibility, as it represents a virtual machine running on the simulating host and can thus be used in context of *any* processor model and *any* kind of application.

*Dynamically scheduled*, compiled simulation reduces simulation time by performing the steps of instruction decoding and operation sequencing prior to simulation. This technique cannot be applied to models using external memories or applications consisting of self-modifying program code.

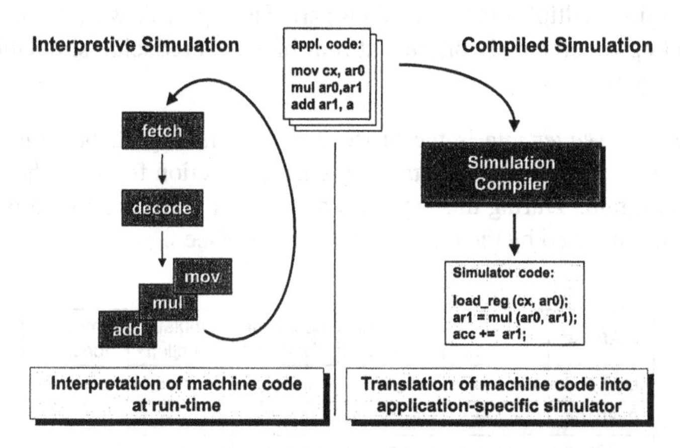


*Figure 7.5.* Interpretive versus compiled simulation.

Besides the compilation steps performed in dynamic scheduling, *static scheduling* and *code translation* additionally implement operation instantiation. While the latter technique is used for instruction-based models, the former is suitable for cycle-based models including instruction pipelines. Beyond, the same restrictions apply as for dynamically scheduled simulation.

The principle of compiled simulation relies on an additional translation step taking place before the simulation is run. This step is performed by a so-called *simulation compiler*, which implements the steps of instruction decoding, operation sequencing, and operation instantiation (see figure 7.5). Obviously, the simulation compiler is a highly architecture-specific tool, which is therefore retargeted from the LISA processor description.

A detailed discussion of the different compiled simulation techniques is given in the following sections, while performance results are given in section 4. The interpretive simulator is not discussed.

## 2.1 Compiled Simulation

The objective of compiled simulation is to reduce the simulation time. Efficient run-time reduction can be achieved by performing repeatedly executed operations only once *before* the actual simulation is run, thus inserting an additional translation step between application load and simulation. The preprocessing of the application code can be split into three major steps.

1 Within the step of *instruction decoding*, instructions, operands, and modes are determined for each instruction word found in the executable object file. In compiled simulation, the instruction decoding is only performed once for each instruction, whereas interpretive simulators decode the same

instruction multiple times, e.g. if it is part of a loop. This way, the instruction decoding is completely omitted at run-time, thus reducing simulation time significantly.

2 *Operation sequencing* is the process of determining all operations to be executed for the accomplishment of each instruction found in the application program. During this step, the program is translated into a table-like structure indexed by the instruction addresses (see figure 7.6).

| address | simulator function | simulator function | simulator function | simulator function | simulator function |
|---|---|---|---|---|---|
| 0x80100 | &sim_func_11 | &sim_func_12 | &sim_func_13 | &sim_func_14 | ... |
| 0x80104 | &sim_func_21 | &sim_func_22 | &sim_func_23 | &sim_func_24 | ... |
| 0x80108 | &sim_func_31 | &sim_func_32 | &sim_func_33 | &sim_func_34 | ... |
| ... | &sim_func_41 | &sim_func_42 | &sim_func_43 | &sim_func_44 | ... |

*Figure 7.6.* Table storing behavioral code to be executed in context of instructions.

The table lines contain pointers to functions representing the behavioral code of the respective LISA operations. Although all involved operations are identified during this step, their temporal execution order is still unknown.

3 The determination of the operation timing (scheduling) is performed within the step of *operation instantiation* and *simulation-loop unfolding*. Here, the behavioral code of the operations is instantiated by generating the respective function calls for each instruction in the application program, thus unfolding the simulation loop that drives the simulation into the next state.

Besides fully compiled simulation, which incorporates all of the above steps, partial implementations of the compiled principle are possible by performing only some of these steps. The accomplishment of each of these steps gives a further run-time reduction, but also requires a non-negligible amount of compilation time. The trade-off between compilation time and simulation time is qualitatively shown in figure 7.7.

There are two levels of compiled simulation which are of particular interest – *dynamic scheduling* and *static scheduling* respectively *code translation*. In case of the dynamic scheduling, the task of selecting operations from overlapping instructions in the pipeline is performed at run-time of the simulation. The static scheduling already schedules the operations at compile-time.

### 2.1.1 Dynamic Scheduling

As shown in figure 7.7, the dynamic scheduling performs instruction decoding and operation sequencing at compile-time. However, the temporal execution

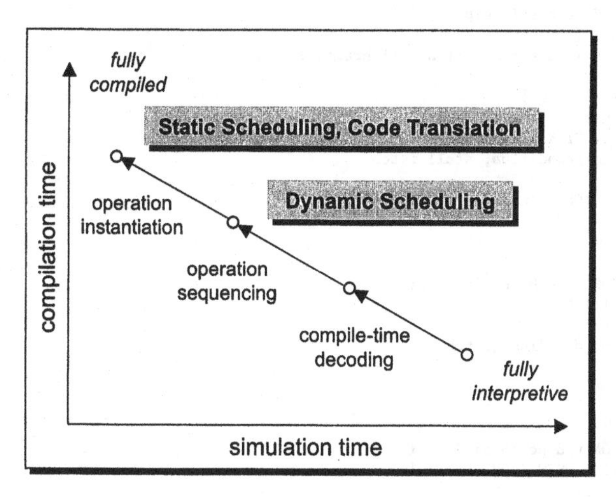


*Figure 7.7.* Levels of compiled simulation.

order of operations is determined at simulator run-time. While the operation scheduling is rather simple for instruction-based models, it becomes a complex task for cycle-based models with instruction pipelines.

In order to exactly reflect the instructions' timing and to consider all possibly occurring pipeline effects like flushes and stalls, a generic pipeline model is employed simulating the instruction pipeline at run-time [167]. The pipeline model is parameterized by the LISA model description and can be controlled via predefined operations. These operations include

- insertion of operations into the pipeline,
- execution of all operations residing in the pipeline,
- pipeline shift,
- removal of operations (flush), and
- halt of entire pipeline or particular stages (stall).

Different from statically scheduled simulation, operations are inserted into and removed from the pipeline dynamically, that means, each operation injects further operations upon its execution. The information about operation timing is provided in the LISA description, i.e. by the activation section as well as the assignment of operations to pipeline stages (cf. chapter 3.2.5 – *timing model*). Example 7.1 shows sample LISA code with the implementation of a conditional branch instruction based on a four-stage pipeline containing a fetch (FE), decode (DC), execute (EX), and write-back (WB) stage. For simplification, the information on coding and syntax is partially omitted in the example.

```
OPERATION decode IN pipe.DC {
  DECLARE {
    GROUP insn = { add || sub || branch };
  }
  CODING { insn }
  IF (insn == branch) THEN {
    ACTIVATION {
      test_condition, stall_fetch
    }
  ACTIVATION { insn }
  }
}

OPERATION branch IN pipe.EX {
  ACTIVATION {
    if (C) {
      upd_pc; flush_fetch
    }
  }
}

OPERATION upd_pc IN pipe.WB {
  BEHAVIOR USES ( OUT PC; ) {
    PC = ...
  }
}
```

*Example 7.1:* Conditional activation of a branch instruction.

Operation *decode* defines a group of three alternative operations *add, sub* and *branch.* The respective operation is activated depending on the coding of the instruction. The activation of an operation leads to the execution of its corresponding behavioral code, however, unlike function calls, execution is delayed until the instruction reaches the pipeline stage the operation is assigned to. Pipeline stage assignment is achieved with the keyword IN followed by pipeline and stage identifier. Assuming the decoding of a branch instruction, the operations *test_condition* and *stall_fetch* are executed immediately in the decode stage (DC), since they are usually not assigned to a particular pipeline stage (not shown). The branch operation, however, is not executed until the occurrence of a pipeline shift, moving the execution context of the instruction into the execute stage. Moreover, the example shows the difference between *compile-time* and *run-time* conditions. While the expression following the capital *IF* can be evaluated upon decoding (operation selection), the *if-else* construct embedded into the activation section must be evaluated during simulator run-time, since it depends on the current state of the processor, i.e. the contents of the condition register *C.* If the condition evaluates true, the operations *flush_fetch* and *upd_pc* are scheduled for execution in the following cycle (when the instruction reaches the write-back stage)[2].

[2]The semicolon inserts an additional delay cycle to ensure that flush_fetch is executed in the following cycle.

Figure 7.8 shows the operation activation chain in case of executing the branch instruction. Operations *main* and *fetch* are not shown in example 7.1.

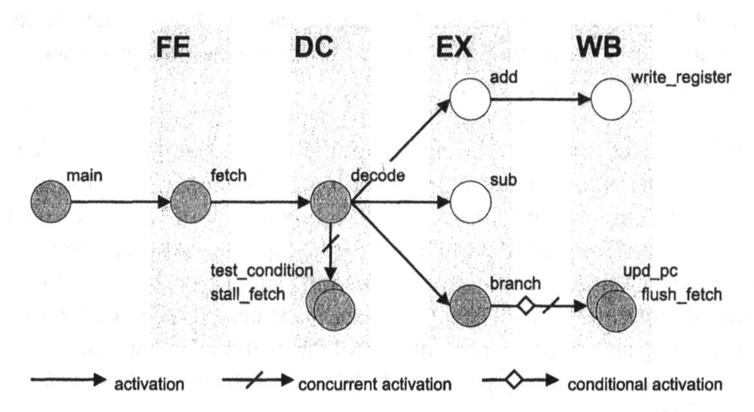


*Figure 7.8.* Operation activation chain for a conditional branch instruction.

It is obvious that the maintenance of the pipeline model at simulation time is expensive. Execution profiling on the generated simulators for the Texas Instruments TMS320C62xx [15] and TMS320C54x [168] revealed that more than fifty percent of the simulator's run-time is consumed by the simulation of the pipeline [120, 169].

The situation can be improved by implementing the step of operation instantiation, consequently superseding the need for pipeline simulation. This, in turn, implies *static* scheduling, in other words, the determination of the operation timing due to overlapping instructions in the pipeline taking place at compile-time. Although there is no pipeline contained in instruction-based processor models, operation instantiation also gives a significant performance increase for these models (cf. section 2.1.3).

### 2.1.2 Static Scheduling

Generally, operation instantiation can be described as the generation of an individual piece of behavioral simulator code for each instruction found in the application program. While this is straightforward for instruction-based processor models, cycle-based, pipelined models require a more sophisticated approach.

Considering instruction-based models, the shortest temporal unit that can be executed is an instruction. That means, the actions to be performed for the execution of an individual instruction are determined by the instruction alone. In the simulation of pipelined models, the granularity is defined by cycles. However, since several instructions might be active at the same time due to overlapping execution, the actions performed during a single cycle are determined by the respective state of the instruction pipeline. As a consequence, instead of instantiating operations for each single instruction of the application program, behavioral code for each occurring *pipeline state* has to be generated. Several of such pipeline states might exist for each instruction depending on the execution context of the instruction, i.e. the instructions executed in the preceding and following cycles.

**Operation Instantiation** The objective of static scheduling is the determination of all possible pipeline states according to the instructions found in the application program. For purely sequential pipeline flow, i.e. in case that no control hazards occur, the determination of the pipeline states can be achieved simply by overlapping consecutive instructions subject to the structure of the pipeline. In order to store the generated pipeline states, *pipeline state tables* are used, providing an intuitive representation of the instruction flow in the pipeline. Inserting instructions into pipeline state tables is referred to as *scheduling* in the following.

A pipeline state table is a two-dimensional array storing pointers to LISA operations. One dimension represents the location within the application, the other the location within the pipeline, i.e. the stage in which the operation is executed. When a new instruction has to be inserted into the state table, both intra-instruction and inter-instruction precedence must be considered to determine the table elements in which the corresponding operations will be entered. Consequently, the actual time an operation is executed at depends on the scheduling of the preceding instruction as well as the scheduling of the operation(s) assigned to the preceding pipeline stage within the current instruction.

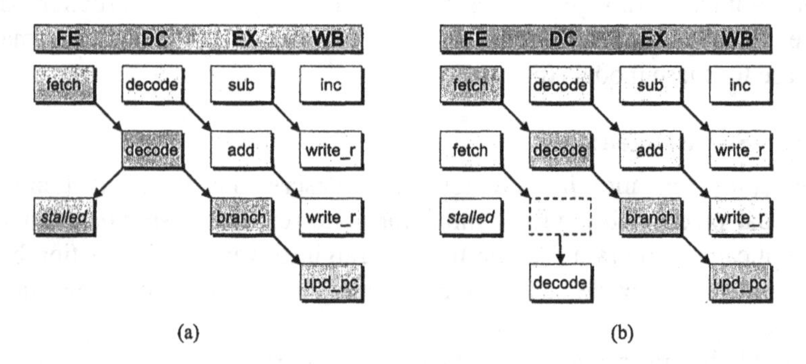


*Figure 7.9.* Inserting operations into the pipeline state table while a stall is issued.

Furthermore, control hazards causing pipeline stalls and/or flushes influence the scheduling of the instruction following the occurrence of the hazard. A simplified illustration of the scheduling process is given in figure 7.9. Here, figure 7.9a shows the pipeline state table after a branch instruction has been inserted, composed of the operations *fetch*, *decode*, *branch*, and *upd_pc* as well as a stall operation. The table columns represent the pipeline stages, the rows represent consecutive cycles (with earlier cycles in upper rows). The arrows indicate activation chains.

The scheduling of a new instruction always follows the intra-instruction precedence, that means, fetch is scheduled before decode, decode before branch, and so on. The appropriate array element for fetch is determined by its assigned pipeline stage (FE) and according to inter-instruction precedences. Since the branch instruction follows the add instruction (which has already been scheduled), the *fetch* operation is inserted below the first operation of add (not shown in figure 7.9a). The other operations are inserted according to their precedences.

The stall of pipeline stage FE, which is issued from the *decode* operation of branch, is processed by tagging the respective table element as stalled. When the next instruction is scheduled, the stall is accounted for by moving the *decode* operation to the next table row respectively next cycle (see figure 7.9b). Pipeline flushes are handled in a similar manner: if a selected table element is marked as flushed, the scheduling of the current instruction is abandoned (see figure 7.10a and 7.10b respectively).

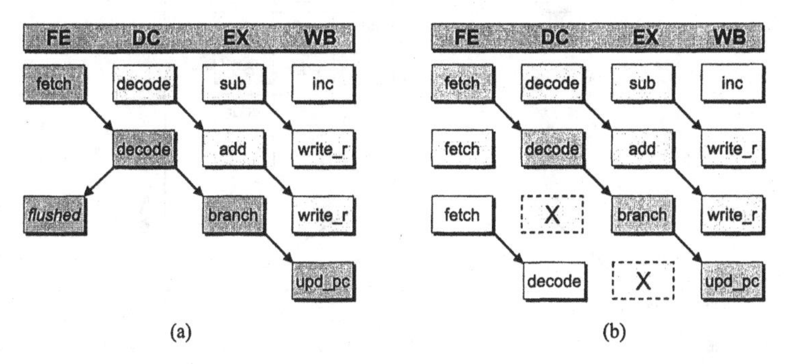


*Figure 7.10.* Inserting instructions into the pipeline state table while a flush is issued.

Assuming purely sequential instruction flow, the task of establishing a pipeline state table for the entire application program is straightforward. However, every sensible application contains a certain amount of control-flow, e.g. loops, interrupting this sequential execution. The occurrence of such control-flow instructions makes the scheduling process at compile-time extremely difficult or in a few cases even impossible.

Generally, all instructions modifying the program counter cause interrupts in the control-flow. Furthermore, only instructions providing an immediate target address – branches and calls whose target address is known at compile-time – can be scheduled statically. If indirect branches or calls occur, it is inevitable to switch back to dynamic scheduling at run-time. However, most control-flow instructions can be scheduled statically. Figure 7.11 exemplarily shows the pipeline states for a conditional branch instruction as found in the TMS320C54x's instruction-set. Since the respective condition cannot be evaluated until the instruction is executed, scheduling has to be performed for both eventualities (condition true respectively false), splitting the program into alternative execution paths. The selection of the appropriate block of pre-scheduled pipeline states is performed by switching among different state tables at simulator run-time. In order to prevent from doubling the entire pipeline state table each time a conditional branch occurs, alternative execution paths are left as soon as an already generated state has been reached. Unless several conditional instructions reside in the pipeline at the same time, these usually have the length of a few rows.

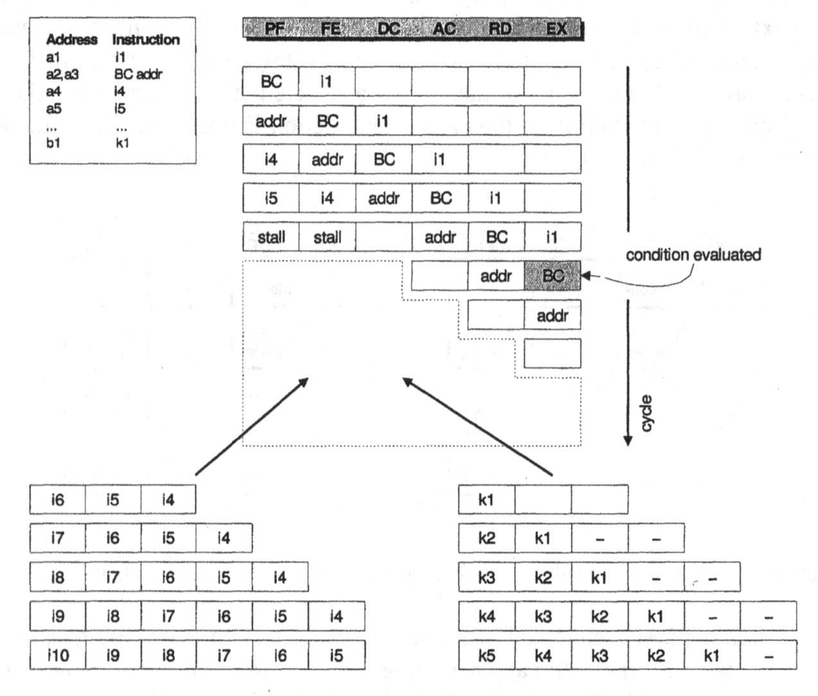


*Figure 7.11.* Pipeline behavior for a conditional branch.

**Simulator Instantiation** After all instructions of the application program have been processed, and thus the entire operation schedule has been established, the simulator code can be instantiated. Example 7.2 shows a simplified excerpt of the generated C code for a branch instruction. Cases represent instructions, while a new line starts a new cycle.

The generated code in example 7.2 corresponds with the pipeline state table shown in figure 7.9b. As the branch instruction causes the stall, address *0x1586*[3] accounts for two cycles in the table.

```
switch (pc) {
  case 0x1584: fetch(); decode(); sub(); inc();
  case 0x1585: fetch(); decode(); add(); write_r();
  case 0x1586: branch(); write_r();
               fetch(); decode(); upd_pc();
  . . .
}
```

*Example 7.2:* Generated simulator code.

### 2.1.3 Code Translation

The need for a scheduling mechanism arises from the presence of an instruction pipeline in the processor model. However, even instruction-based processor models without pipeline benefit from the step of operation instantiation. The technique applied here is called *instruction-based code translation.* Due to the absence of instruction overlap, simulator code can be instantiated for each instruction independently, thus simplifying simulator generation to the concatenation of the respective behavioral code specified in the LISA description. In contrast to direct binary-to-binary translation techniques [170], the translation of target-specific into host-specific machine code uses C-source-code as intermediate format. This keeps the simulator portable, and thus independent from the simulation host.

Since the instruction-based code translation generates program code that linearly increases in size with the number of instructions in the application, the use of this simulation technique is restricted to small and medium sized applications (less than $\approx 10k$ instructions, depending on model complexity). For large applications, the resultant worse cache utilization on the simulation host reduces the performance of the simulator significantly (cf. section 4).

## 3. Debugging

The simulation run can be controlled and visualized by employing a graphical debugger frontend. As the LISA environment allows simulator retargeting to arbitrary architectures, the debugger also needs to be adaptable. This concerns the placement of processor resources like registers, I/O ports, and memories into the respective windows of the debugger frontend.

Today, debuggers for embedded processors are only available for target-specific tools that are tightly coupled to the processor simulator. Furthermore, debuggers for high-level languages like gdb [171] and its graphical frontend

[3] Address *0x1586* is the fetch address of the branch instruction.

ddd [172] do not provide target-specific, micro-architectural information like pipeline states at all.

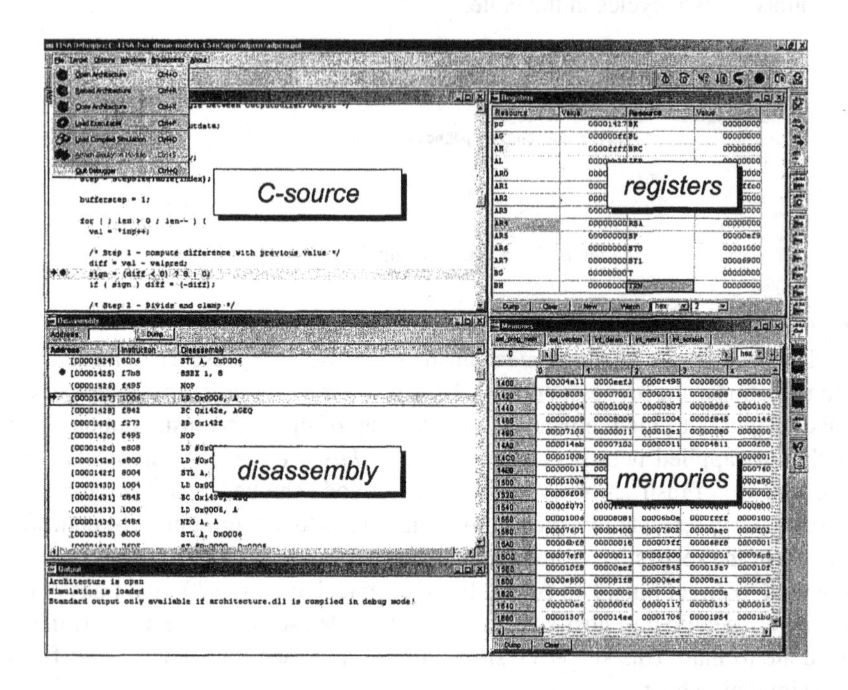


*Figure 7.12.* Graphical LISA debugger frontend which visualizes and controls the simulation.

Figure 7.12 shows the LISA debugger running an application on the Texas Instruments TMS320C54x DSP. The debugger windows are parameterized by the information of the underlying LISA model (cf. chapter 3.2.1 – *memory model*).

## 4. Case Studies

To examine the quality of the generated software development tools, four different architectures have been considered. The architectures were carefully chosen to cover a broad range of architectural characteristics and are widely used in the field of digital signal processing (DSP) and micro-controllers ($\mu$C). Moreover, the abstraction level of the models ranges from cycle-based (TMS320C62x) to instruction-based (ARM7100).

- **ARM7100.** The ARM7 family is a range of low-power 32-bit RISC microprocessor cores of Advanced RISC Machines Ltd. (ARM) [16] optimized for cost and power-sensitive consumer applications. Areas of application comprise personal audio (MP3, WMA, AAC players), basic wireless handset, pager, digital still camera, and PDA. The realization of an instruction-based LISA model of the ARM7100 took approximately two weeks. The model comprises 4700 lines of LISA- and C-code and was validated against the ARMulator [166] of ARM.

- **ADSP2101** The ADSP-21xx family processors are 16-bit fixed-point single-chip micro-computers of Analog Devices optimized for digital signal processing (DSP) and other high speed numeric processing algorithms [173]. Primary area of application is automotive, especially motor control. The realization of a cycle-based LISA model of the ADSP2101 took approximately three weeks. The model comprises 6700 lines of LISA- and C-code and was validated against the VisualDSP++ simulator [164] of Analog Devices.

- **TMS320C541** The TMS320C54x family is a range of high performance 16-bit fixed-point digital signal processors (DSP) [168] of Texas Instruments. The architecture contains dedicated, application-specific instructions working on a compare, select, and store unit (CSSU) for the add/compare selection of the viterbi operator. Primary area of application is mobile communication, especially GSM. The realization of a cycle-based model (including pipeline behavior) took approximately eight weeks. The model comprises 16000 lines of LISA- and C-code and was validated against the Code Composer Studio (CCS) simulator [165] of Texas Instruments.

- **TMS320C6201** The TMS320C62x family is a general-purpose 256-bit fixed-point DSP of Texas Instruments [15]. It is based on a high-performance, advanced very-long-instruction-word (VLIW) architecture (VelociTI) developed by Texas Instruments. Primary area of application are multichannel and multifunction applications, especially in the area of base-stations for mobile communication. The realization of a cycle-based model (including pipeline behavior) took approximately six weeks. The model comprises 11600 lines of LISA- and C-code and was validated against the Code Composer Studio (CCS) simulator [165] of Texas Instruments.

All of the above mentioned architectures were modeled by one designer on the respective abstraction level with LISA and software development tools were generated successfully. The speed of the generated tools was then compared with the tools shipped by the respective architecture vendor. In order to provide comparable conditions, the LISA tools are working on the same level of accuracy as the vendor tools. The vendor tools are exclusively using the interpretive simulation technique. The LISA models have been validated against the vendor tools by running a mix of applications taken from the DSPStone benchmarking suite [174] on both simulators and dumping the state (register-set and memory) of the processor after each control step.

## 4.1 Efficiency of the Generated Tools

Measurements took place on an AMD Athlon system with a clock frequency of 1300 MHz. The system is equipped with 768 MB of RAM and is part of the networking system. It runs under the operating system Windows 2000. Tool compilation was performed with Visual C++, version 6.0, Microsoft Inc. [175]. Time was measured with the *clock()* function which is part of the ANSI-C standard [176] and calculates the processor time of the calling process.

The set of simulated applications on the respective architectures comprises a simple 20 tap FIR filter, an ADPCM transcoder, and a GSM full-rate speech codec.

The ADPCM G.721 (Adaptive Differential Pulse Code Modulation) standard operates at 32 kbits/sec and is one of the oldest speech coding standards which plays an important role even in recent designs such as DECT (Digital Enhanced Cordless Telecommunications). It is specified up to bit-accurate test sequences provided by the International Telecommunication Union (ITU) [177]. GSM, the Global System for Mobile Communications, is a digital cellular communications system which has rapidly gained acceptance and market share worldwide, although it was initially developed in a European context. The GSM full-rate speech codec operates at 13 kbits/s and uses a Regular Pulse Excited (RPE) codec. It is specified up to bit-accurate test sequences provided by the European Telecommunication Standardization Institute (ETSI) [178]. For both ADPCM and GSM codec, the software simulators process the respective test sequences recommended by the standard. Instead of running the GSM codec, the ARM7100 processes an ATM-QFC protocol application [179], which is responsible for flow-control and configuration in an ATM port-processor chip [180].

### 4.1.1 Advanced Risc Machines ARM7100 $\mu$C

As indicated in section 4, a LISA model was realized for the ARM7100 architecture at instruction-set accuracy. The model is derived from the information on the architecture that can be found on the publicly accessible web-pages of Advanced Risc Machines [16].

The compilation time of the ARM7100 LISA model by the LISA processor compiler *lc* is shown in table 7.1. Besides, the compilation times of the generated C++ code for assembler, linker, disassembler, and software simulator are presented.

The generated software simulator can simulate the target application with different simulation techniques – interpretive and compiled (cf. section 2.1). As explained, the application of the compiled simulation principle requires an additional step to be performed before the simulation is run – simulation compilation. Table 7.2 shows the compilation times for different applications under

*Table 7.1.* Compilation time of ARM7100 model and the generated software tools.

| | *LISA model* | *Assembler* | *Linker* | *Disassembler* | *Simulator* |
|---|---|---|---|---|---|
| *Compile-Time* | 4.1 sec | 4.5 sec | 4.5 sec | 2.5 sec | 5.1 sec |

test in conjunction with the simulation techniques requiring compilation – compiled simulation with dynamic scheduling and code translation. Lines of code refers to the size of the application, not to the number of executed instructions. As expected, simulation compilation time increases with the number of lines of code of the target application.

*Table 7.2.* Simulation compilation times for different applications running on the ARM7100.

| | *Lines of Code* | *Dynamic Scheduling* | *Code Translation* |
|---|---|---|---|
| *FIR* | 59 | 0.1 sec | 15 sec |
| *ADPCM* | 460 | 0.5 sec | 123 sec |
| *ATM-QFC* | 12621 | 0.9 sec | 612 sec |

Figure 7.13 shows the simulation speed of the applications under test on the vendor simulator (ARMulator) and the generated simulator. The ARMulator works with the interpretive simulation technique. The LISA simulator is benchmarked in interpretive and compiled mode (dynamic scheduling and instruction-based code translation). The interpretive simulator distributed by ARM Ltd. runs at a speed of approximately three MIPS (mega-instructions per second). The generated simulator running in interpretive mode achieves approximately two MIPS, while run in compiled mode with dynamic scheduling it performs at approximately eight MIPS. As expected, the speed of the simulator varies only negligibly with the application under test.

However, for instruction-based code translation, this does not hold true. Depending on the size of the application, the simulation speed ranges from 58 MIPS to only 18 MIPS. This is due to the fact that the highest degree of simulation compilation (cf. section 2.1.3) requires additionally the step of operation instantiation to be performed. Naturally, operation instantiation leads to large executable simulations, which increase in size proportionally with the application under investigation. As the executable simulator becomes bigger, cache effects play an increasing role in simulation performance on the simulating host. Therefore, simulation speed declines with bigger applications.

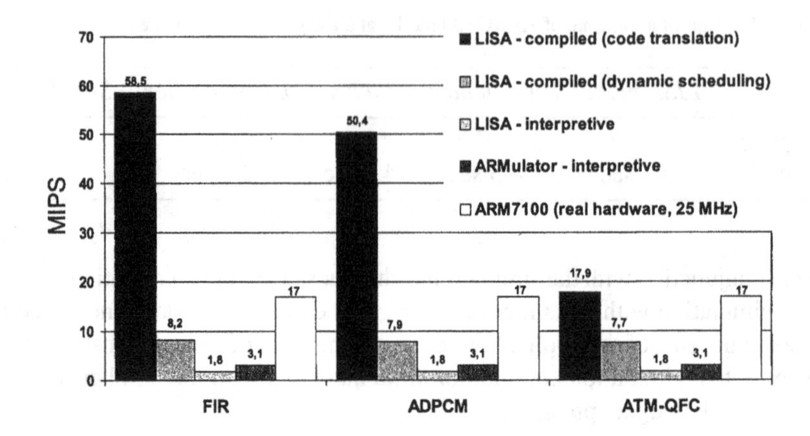


*Figure 7.13.* Simulation speed of the ARM7100 $\mu$C at instruction accuracy.

As indicated in section 2.1.3, instruction-based code translation is most useful for small and medium size applications.

It is interesting to note the performance of the real architecture running at a clock frequency of 25 MHz. This translates to an average instruction throughput of approximately 17 MIPS. Using a quick host machine (as in this case study), the software simulator even outperforms the real hardware. This allows early software development and rapid prototyping before the real silicon is at hand.

### 4.1.2 Analog Devices ADSP2101 DSP

The LISA model of the ADSP2101 architecture of Analog Devices was realized at cycle-accuracy. It is derived from information on the architecture that can be found in the architectural manual [173].

*Table 7.3.* Compilation time of ADSP2101 model and the generated software tools.

| | *LISA model* | *Assembler* | *Linker* | *Disassembler* | *Simulator* |
|---|---|---|---|---|---|
| *Compile-Time* | 2.2 sec | 2.2 sec | 2.4 sec | 4.7 sec | 7.9 sec |

The compilation time of the ADSP2101 LISA model by the LISA processor compiler *lc* is shown in table 7.3. Besides, the compilation times of the generated C++ code of assembler, linker, disassembler, and software simulator are presented.

As the ADSP2101 contains neither an instruction nor a data pipeline, the instruction-based code translation technique is applied in simulation even though the model is cycle-based. Table 7.4 shows the simulation compilation times for different applications under test in conjunction with the simulation techniques requiring compilation – compiled simulation with dynamic scheduling and code translation.

*Table 7.4.* Simulation compilation times for different applications running on the ADSP2101.

| | *Lines of Code* | *Dynamic Scheduling* | *Code Translation* |
|---|---|---|---|
| *FIR* | 58 | 0.1 sec | 22 sec |
| *ADPCM* | 985 | 0.3 sec | 85 sec |
| *GSM* | 12328 | 0.5 sec | 756 sec |

Figure 7.14 shows the simulation speed of the applications under test on the vendor simulator (VisualDSP++) and the generated simulator. The VisualDSP++ simulator of Analog Devices works with the interpretive simulation technique. The LISA simulator is benchmarked in interpretive and compiled mode (dynamic scheduling and instruction-based code translation).

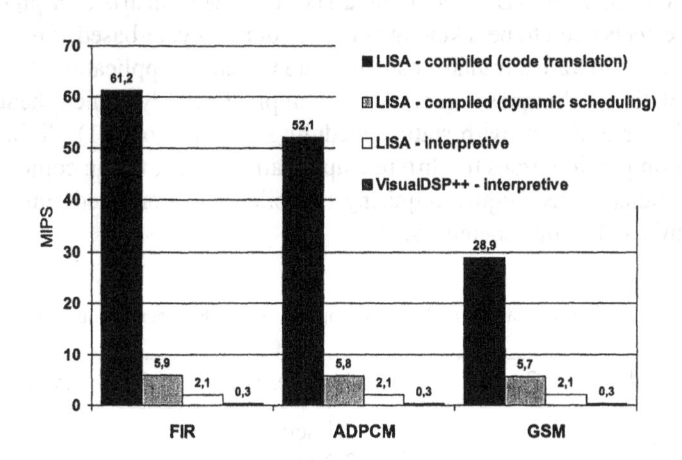


*Figure 7.14.* Simulation speed of the ADSP2101 DSP at cycle-accuracy.

The interpretive simulator distributed by Analog Devices runs at a speed of approximately 300 KIPS (kilo-instructions per second). The generated simulator running in interpretive mode achieves approximately two MIPS, while run in compiled mode with dynamic scheduling it performs at approximately nine MIPS. The instruction-based code translation technique speeds up simulation to values between 61 MIPS and 29 MIPS depending on the size of the application under test. The reason for this are again cache effects on the simulating host machine that increase with the size of the application under test.

### 4.1.3 Texas Instruments TMS320C541 DSP

The LISA model of the TMS320C541 architecture of Texas Instruments was realized at cycle-accuracy. It is derived from the information on the architecture that can be found in the architectural manual [168].

The compilation time of the TMS320C541 LISA model by the LISA processor compiler *lc* is shown in table 7.5. In addition, the compilation times of the generated C++ code of assembler, linker, disassembler, and software simulator are presented.

*Table 7.5.* Compilation time of the C541 model and the generated software development tools.

| | *LISA model* | *Assembler* | *Linker* | *Disassembler* | *Simulator* |
|---|---|---|---|---|---|
| *Compile-Time* | 4.2 sec | 2.7 sec | 2.9 sec | 7.3 sec | 16.4 sec |

As the TMS320C541 DSP contains a six stage deep instruction pipeline, pipelining effects need to be taken into account during cycle-based simulation. The generated software simulator can simulate the target application with different simulation techniques – interpretive, compiled with dynamic scheduling and compiled simulation with static scheduling (cf. section 2.1). Table 7.6 shows the compilation times for different applications under test in conjunction with the simulation techniques requiring compilation – compiled simulation with dynamic and static scheduling.

*Table 7.6.* Simulation compilation times for the applications under test running on the C541.

| | *Lines of Code* | *Dynamic Scheduling* | *Static Scheduling* |
|---|---|---|---|
| *FIR* | 51 | 0.1 sec | 5 sec |
| *ADPCM* | 360 | 0.3 sec | 38 sec |
| *GSM* | 4601 | 0.6 sec | 288 sec |

Figure 7.15 shows the simulation speed of the applications under test on the vendor simulator (Code Composer Studio with C541 simulator backend) and the generated simulator. The Code Composer Studio simulator of Texas Instruments works with the interpretive simulation technique. The LISA sim-

ulator is benchmarked in interpretive and compiled mode (dynamic and static scheduling).

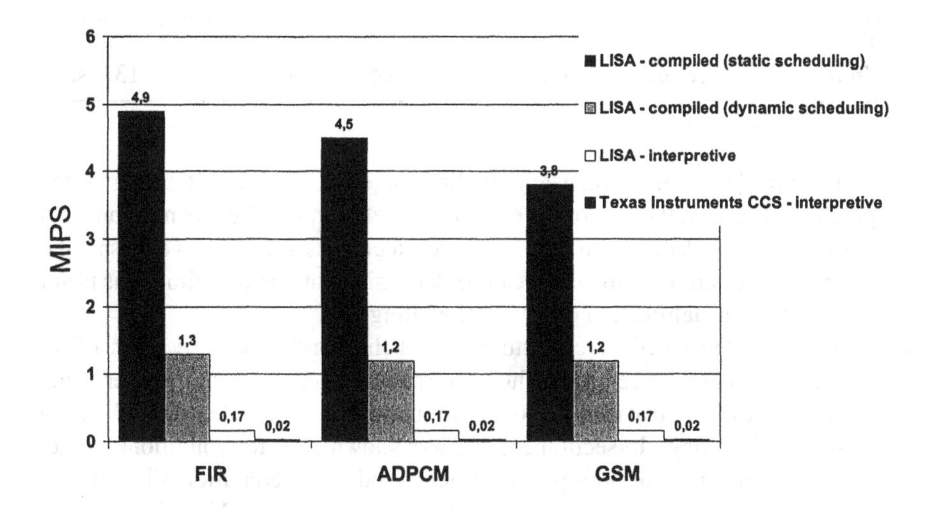


*Figure 7.15.* Simulation speed of the TMS320C541 DSP at cycle-accuracy.

The interpretive simulator distributed by Texas Instruments runs at a speed of approximately 30 KIPS. The generated simulator running in interpretive mode achieves approximately 170 KIPS, while run in compiled mode with dynamic scheduling it performs at approximately 1,3 MIPS. Compiled simulation with static scheduling speeds up simulation to values between 4,9 MIPS and 3,8 MIPS depending on the size of the application under test. The cache effects due to the size of the executable application are visible, however, do not have such a strong impact on simulation performance as for instruction-based code translation.

### 4.1.4 Texas Instruments TMS320C62x DSP

The LISA model of the TMS320C6201 architecture of Texas Instruments was realized at cycle-accuracy. It is derived from the information on the architecture that can be found in the architectural manual [15].

The compilation time of the C6201 LISA model by the LISA processor compiler *lc* is shown in table 7.7. Besides, the compilation times of the generated C++ code of assembler, linker, disassembler, and software simulator are presented.

The C6201 DSP architecture is based on a superscalar instruction dispatching mechanism. Here, 256-bit wide VLIW instruction word is fetched from memory and split up into several execute packets. These are issued in con-

*Table 7.7.* Compilation time of C6201 model and the generated software development tools.

| | *LISA model* | *Assembler* | *Linker* | *Disassembler* | *Simulator* |
|---|---|---|---|---|---|
| *Compile-Time* | 8.1 sec | 6.5 sec | 6.5 sec | 5.1 sec | 13.0 sec |

secutive cycles from the dispatch pipeline stage into the execute stages of the pipeline. Thereby, the size of the execute packets can vary between 32-bits and 256-bits. As the decision on the size of the execute packet is based on a specific bit in the instruction word which can be determined at compile-time, this is not a factor for disqualification of static scheduling.

However, the C6201 architecture allows the conditional execution of all instructions based on register values (fully predicated). As the register value can only be evaluated at run-time of the simulation, this excludes the application of static scheduling. In section 2.1.2 it was shown, that for conditional direct branches schedules for both paths are generated. As a conditional instruction basically represents such a conditional direct branch and one VLIW instruction word in the C6201 architecture consists of 8 conditional micro-instructions, it would be required to generate $2^8 = 256$ alternative pathes for just one VLIW instruction. For this reason static scheduling was not applied in this case.

The generated software simulator can simulate the target application with interpretive and dynamically scheduled compiled simulation. Table 7.8 shows the compilation times for compiled simulation with dynamic scheduling.

*Table 7.8.* Simulation compilation times for the applications under test running on the C6201.

| | *Lines of Code* | *Dynamic Scheduling* |
|---|---|---|
| *FIR* | 61 | 0.1 sec |
| *ADPCM* | 461 | 0.2 sec |
| *GSM* | 23312 | 0.5 sec |

Figure 7.16 shows the simulation speed of the applications under test on the vendor simulator (Code Composer Studio with C6201 simulator backend) and the generated simulator. The Code Composer Studio simulator of Texas Instruments works with the interpretive simulation technique. The LISA simulator is benchmarked in interpretive and compiled mode (dynamic scheduling).

The interpretive simulator distributed by Texas Instruments runs at a speed of approximately 25 KIPS. The generated simulator running in interpretive mode

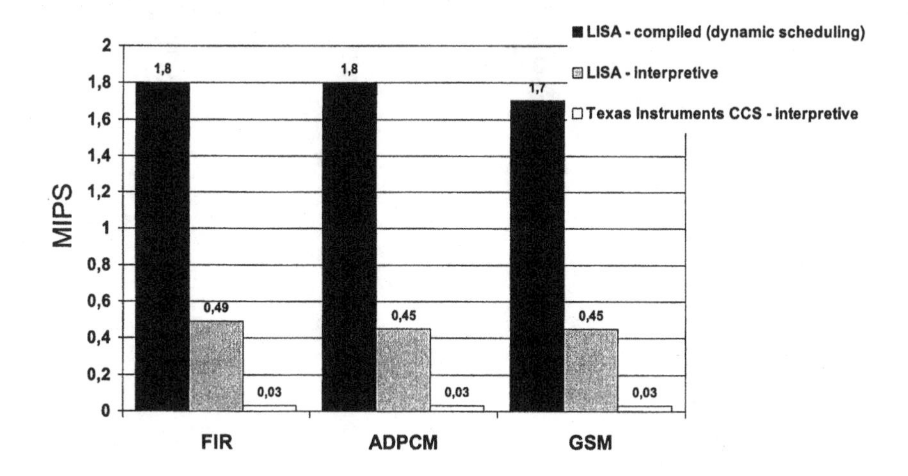


*Figure 7.16.* Simulation speed of the TMS320C6201 DSP at cycle-accuracy.

achieves approximately 0,5 MIPS, while run in compiled mode with dynamic scheduling it performs at approximately 1,8 MIPS.

## 5. Concluding Remarks

In this chapter the third quadrant of the LISA processor design platform was introduced – software development tools for application design. In general, the feasibility of generating a set of *production quality* software development tools from machine descriptions in the LISA language was shown. The suite of software development tools comprises assembler, linker, simulator, disassembler, and graphical debugger. The simulator is enhanced in speed by applying the compiled simulation principle – where applicable – which boosts simulation speed by one to two orders in magnitude. In three case studies using real-world processor architectures (ARM7100, ADSP2101, C54x, and C62x), benchmarking results showed the achieved simulation performance compared to the tools provided by the respective architecture vendor. A complete list of architecture models realized with LISA can be found in [181].

Figure [illegible] [illegible] [illegible]

[illegible] more [illegible] and [illegible] [illegible]

# 5 [illegible] Concluding Remarks

In this chapter the [illegible] [illegible] [illegible] [illegible] [illegible] [illegible] [illegible] [illegible] [illegible] [illegible] [illegible] [illegible] where [illegible] [illegible] [illegible] [illegible] [illegible] [illegible] [illegible] [illegible] [illegible] [illegible] [illegible] model [illegible] [illegible]

Chapter 8

# SYSTEM INTEGRATION AND VERIFICATION

Today, typical single chip electronic system implementations include a mixture of $\mu$Cs, DSPs as well as shared memory, dedicated logic (ASICs), and interconnect components [182]. To enable designers to create such complex systems, new system design methodologies have been introduced in recent years. One example is the orthogonalization of concerns [140], i.e. the separation of various aspects of design to allow a more effective exploration of alternative solutions and to ease verification. Here, it is proposed to separate between:

- function (what the system is supposed to do) and architecture (how it does it), and
- communication and computation.

The mapping of function to architecture is the essential step from conception to implementation. Here, the overall function has to be partitioned into hardware and software. The software portion is then further partitioned into processor classes like ASIP, DSP, $\mu$C, etc. The motivation behind this decision can be performance of the application on a particular processor or the need for flexibility and adaptivity.

This chapter deals with the fourth quadrant of the LISA processor design platform – system integration and verification. It will be shown that processor models generated from LISA machine descriptions can be easily integrated into system simulation environments on various levels of abstraction, by this coping with the proposed separation between function and architecture. For this purpose, the LISA processor design platform generates interfaces to the software simulators that allow instantiation and steering of processor models in context of arbitrary system simulation environments. Besides, systems often comprise a variety of processors that need to be observed during verification. It is obvious that it is not feasible to manage several processor cores running in different debuggers at once. For this reason, the LISA simulator debugger has been enhanced to administrate multiple processor cores. The chapter will close with a case study showing the integration of LISA processor simulators into the commercially available CoCentric System Studio (CCSS) environment of Synopsys.

## 1. Platform-Based Design

Platform-based designs have recently become a common term in embedded systems for today's electronic products. What it means is basically the creation and use of a system architecture consisting of programmable processor cores, I/O subsystems, and memories. The idea behind this is to find common architectures that can support a variety of applications as well as future evolutions of a given application to reduce design costs. Consequently, this approach implies a longer time-in-market.

A family of micro-architectures that allow substantial reuse of software is called *hardware platform.* This hardware platform is then abstracted to a level where the application software sees only a high-level interface of the hardware – the application programming interface (API). Usually, a software layer is used to perform this abstraction. This layer wraps different parts of the hardware platform to hide them from the software designer: the programmable cores and memory subsystem via operating systems, the I/O subsystem via device drivers, and the network connection via a network communication subsystem. This abstraction layer is called the *software platform.* The combination of hardware and software platform is frequently called the *system platform.*

Summarizing, the development of a system platform involves understanding the application domain, developing an architecture and a micro-architecture that is specialized to that application domain in cadence with the development of the software development tools to program the specialized architecture. The basic idea of system platforms and platform-based design is depicted in figure 8.1 (by K. Keutzer *et al.* [140]).

The vertex of the two cones represents the API or programmer's model, i.e. the abstraction of the hardware platform. When a system is designed, the system designer maps the target application into an abstract representation that addresses a family of platforms which can be chosen from while optimizing cost, efficiency, energy consumption, and flexibility. Obviously, there is a trade-off between the level of abstraction of the programmers' model and the number and diversity of platform instances covered. The more abstract the programmers' model the richer the set of platform instances but also the more difficult is it to find the *optimal* platform instance.

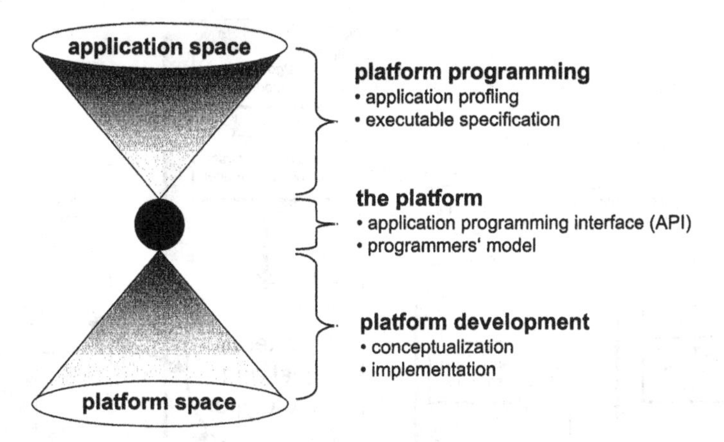


*Figure 8.1.* Platform-based design – abstraction and design flow.

## 2. Enabling Platform-Based Design

The emerging platform-based design paradigm (cf. section 1) poses enormous challenges for the designer to conceptualize, implement, verify, and program a platform which provides sufficient performance, flexibility, and power efficiency as required by application domains like wireless communications and broadband networking.

The envisioned LISA-based design and verification flow of such complex, heterogeneous platforms is depicted in figure 8.2. The proposed platform design methodology can be subdivided into three different phases: platform conceptualization, platform design and verification, and platform programming. These phases will be illuminated in the following.

**Platform conceptualization.** Platform design starts with the definition and extensive profiling of the algorithms representing the target application domain to identify performance bottlenecks. This task is typically performed with algorithm design tools like CCSS [26] for wireless communication or OPNET [142] for networking. In a next step, the algorithm model is wrapped with SystemC [183] modules, which are annotated with timing budgets to create an abstract architecture model. SystemC is a modeling platform consisting of C++ class libraries and a simulation kernel for design on the system-behavioral and register-transfer-levels. Depending on the complexity and application domain, either cycle-based transaction level modeling (TLM) [184] or the more abstract GRACE++ channel library [185] can be employed to build the abstract architecture model. Profiling of the abstract architecture guides the definition of the system architecture and the coarse partitioning into dedicated hardware, programmable hardware, and software. The abstract architecture model also

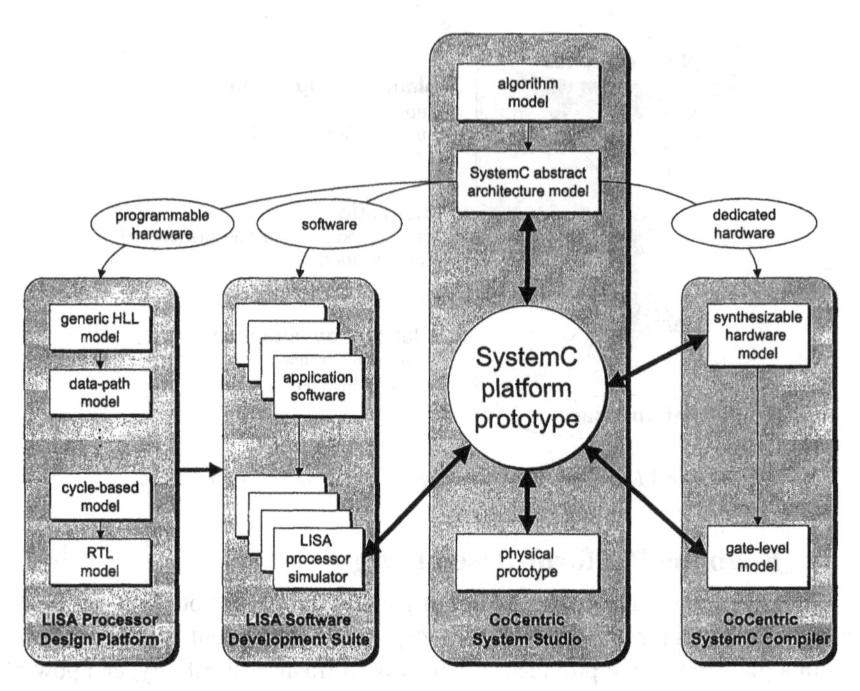


*Figure 8.2.* Platform conceptualization, design, and verification.

serves as a *virtual prototype* of the platform by providing capabilities to integrate hardware and software implementation models of subsequent refinement stages for detailed profiling and functional verification.

**Platform design and verification.** Micro-architecture exploration and implementation of dedicated and programmable hardware modules are carried out within the respective design environments. On the software side, application-specific processor design is performed using the tools provided by the LISA processor design platform (cf. chapter 5). Starting on an application-centric level of abstraction, the processor model is refined in cadence with the application to the level of implementation. On each level of abstraction, the generated software simulators can be seamlessly integrated into the system via the simulators' co-simulation API (cf. section 3). On the hardware side, the CoCentric SystemC compiler [186] enables SystemC-based architecture exploration and implementation of dedicated hardware components. It is important that for functional verification and profiling all implemented components – both hardware and software – can be integrated into a SystemC simulation forming a *virtual prototype* at any time in the design process. The different levels of abstraction can be adjusted by the GRACE++ abstraction interface [187] which allows coupling of both LISA software simulators, HDL simulators, and pure C/C++ code on arbitrary levels of abstraction via their respective co-simulation interfaces.

**Platform programming.** During the post-production phase, the platform is adapted to the actual application by programming the integrated processor cores and configuring the dedicated hardware blocks. For this purpose, the LISA software simulators and debuggers in cadence with the profiling capabilities of the LISA tools provide the software centric view of the mixed SystemC-LISA platform prototype.

# 3. Software Simulator Integration

In order to support the system integration and verification, the software simulators generated from LISA processor models provide a well-defined API to easily interconnect with other simulators. This API controls the simulators by stepping, running, and setting breakpoints in the application code and by providing access to the processor resources.

Besides the need for a flexible API to integrate the software simulators, the overall system simulation performance is a key factor for a successful design. It is obvious that for the verification and evaluation of the complete system, large test-vectors are required. According to R. Camposano [188], the amount of test-vectors needed for verification rises by a factor of 100 every six years, which is ten times the increase of the number of gates on a chip as stated by Moore's law. This clearly indicates that the system simulation speed is crucial when designing complex systems.

Thirdly, due to the heterogeneity of complex systems, the debug ability and thus visibility of the respective components are key factors to enable efficient profiling and verification. Considering only the software portion of a system composed of a variety of processor cores it becomes clear that having separate debuggers for all processors will not contribute to the clearness of the system.

## 3.1 Application Programming Interface (API)

APIs are designed for universal control and visibility of the linked element. They hide details of the target by providing well defined methods to read and modify the state. This is the key for IP protection as only the interfaces are visible to the end-user, while providing the target itself as a *black box* (e.g. a library).

The API generated from LISA processor models remains the same independently of the abstraction level of the underlying model. Among others, it provides the following functions to instantiate and steer the simulator:

- model constructor and destructor,

- load application,
- reset processor,
- advance one control step,
- set or clear breakpoint at specific address, and
- run model until breakpoint is hit.

The model constructor is called to create one instance of the simulator. At this time, memory is allocated for processor resources defined in the resource section of the underlying LISA processor model, which are global by definition. In order to run an application, it has to be loaded first. At the time the application is loaded, the processor is automatically reset and can be pushed forward by using one of the API functions. Here, a control step is referred to as executing the reserved operation *main* in the LISA processor model one time. In case of executing the target application on a generic HLL model, this corresponds to executing one HLL statement within the application while in an RTL micro-architectural model this equals the execution of one cycle on the target architecture (cf. chapter 5). Besides these, there are further functions available to chose e.g. a certain simulation technique, set and clear watch-points, and access profiling information.

Furthermore, the API provides functions which are model specific. These functions concern the processor resources that can be read and written from outside the model. In order to generate the respective API functions for a particular resource, the resource has to be labelled with the keyword PIN in the underlying LISA processor description (cf. chapter 3.2.1 – *memory model*). In case of a generic HLL LISA model, these resources will be variables that require access from outside the processor model while in an RTL micro-architectural model they will represent the pins of the real processor architecture.

Example 8.1 shows sample C++ code with the instantiation of two processor models. The processors in the example are derived from LISA models of the ARM7 $\mu$C and the C54x DSP described in chapter 7. On the ARM7, a simple operating system is executed while the C54x runs a GSM codec. This is a typical setup found in many mobile devices for GSM. The models are pushed forward in a loop until one processor hits a breakpoint. For simplification, the setting of a breakpoint in the application of either one of the two processors is not shown in the example. The processors are communicating via the pins *in_pin* and *out_pin*.

## 3.2 Dynamic Debugger Connection

The processor models instantiated in example 8.1 can be used to debug and verify the behavior of the respective processor on its pins. However, in order

```
#include <ARM7_cosim_API.gen.h>
#include <C54x_cosim_API.gen.h>

int main {
  ARM7_cosim_api *ARM_processor = 0;
  C54x_cosim_api *C54x_processor = 0;

  ARM_processor = new ARM7_cosim_api();                   // Instantiate ARM7 model
  ARM_processor->load("op_sys.out");                      // Load Application on ARM7
  C54x_processor = new C54x_cosim_api();                  // Instantiate C54x model
  C54x_processor->load("gsm.out");                        // Load Application on C54x

  // Simulators are run in loop
  while(!ARM_processor->breakpoint_hit() || !C54x_processor->breakpoint_hit()) {
     ARM_processor->advance_one_control_step();            // Step ARM7
     C54x_processor->advance_one_control_step();           // Step C54x

     C54x_processor->in_pin = ARM_processor->out_pin;      // Exchange values of pins
  }

  delete ARM_processor;
  delete C54x_processor;
}
```

*Example 8.1:* Instantiation of two LISA processor simulators in a C++ program.

to get full visibility of all internal states including registers, memories, and the disassembled instructions of the application to be executed, a graphical debugger is required.

To enable the usage of e.g. processor debuggers in a system simulation environment, tools are required that glue together heterogeneous components via a *simulation backplane*. Today, such hardware-software co-simulation environments are offered commercially by companies like Mentor Graphics (*Seamless* [128]), and Synopsys (*Eaglei* [130]). The underlying concept of all of the above mentioned commercial environments to set up the simulation backplane is similar: both, processor simulators, HDL simulators, and plain C/C++ code with their respective graphical debuggers run in separated processes on the simulating host and communicate via an inter-process communication (IPC) mechanism [189].

Figure 8.3 shows simulation coupling via IPC connection. Here, the graphical debugger together with the processor simulator core are running in one process on the simulating host while the system instantiating the debugger runs in a separate process. The whole communication between the system simulation environment and the processor – both control and data – is going via the IPC connection. However, it is a well known fact that this type of communication is slow. The reason for this is that a protocol has to be used between the two processes to communicate in both directions control information and new data that is written to and read from the processor pins. As conservative synchronization is used between the simulators, i.e. data is exchanged after every control step, simulation speed is limited by the speed of the IPC connection. In the past, the

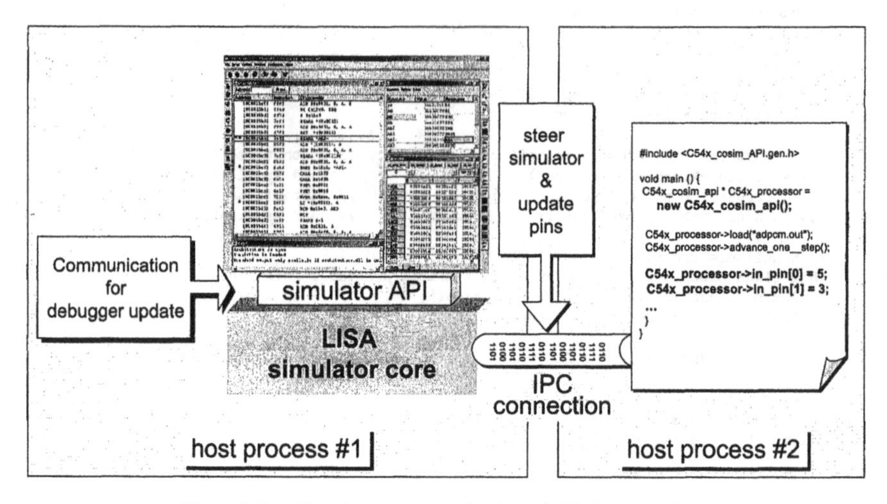


*Figure 8.3.* Simulator communication via IPC connection.

simulation slowdown due to this communication mechanism was annoying but acceptable, as the cycle-based hardware simulators integrated into the system were the limiting factor for overall system simulation speed [32].

However, with the advent of SystemC as a system modeling language that can also be used for synthesis [186] things have changed significantly. Now, the whole system consisting of heterogeneous components can be modeled and simulated using C/C++. Due to SystemC 2.0's capability to model the system on transaction level [184] while staying cycle-based in the domain of time, system simulation speed is increased enormously.

To cope with this, the system integrator of the LISA processor design platform decouples the simulator core, which is generated as C++ code from the underlying LISA processor model, from the graphical debugger frontend. Figure 8.4 displays the principle.

Here, the simulator core containing the behavioral code to be executed, the instruction decoder, the processor resources, and the simulator control including breakpoint management are linked as a library to the system simulator. This is feasible as both the system as well as the software simulator are based on the language C/C++. The synchronization between the simulators is still conservative, however, communication concerning control and data between the system simulator and the LISA processor simulator is now taking place on the basis of pure function calls. By this, the simulation bottleneck of the IPC connection is eliminated.

The communication between the simulator core and the graphical debugger frontend is established via a TCP/IP connection. This connection can be set up dynamically at run-time of the system simulation. The data sent via the

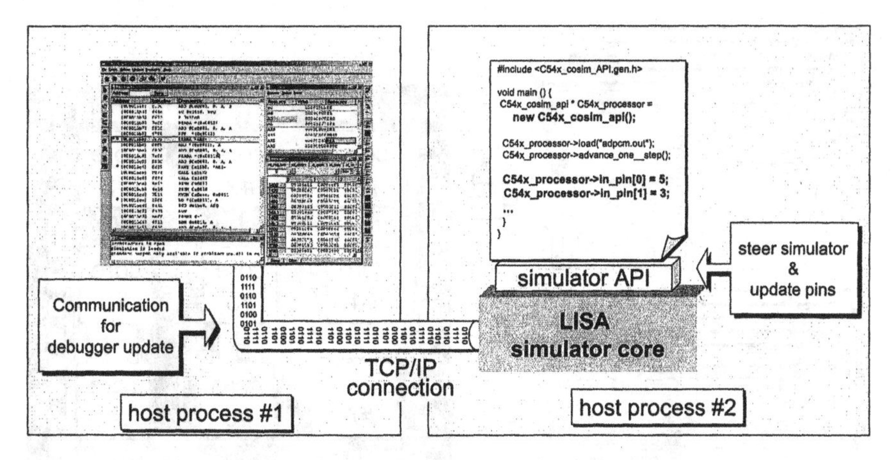


*Figure 8.4.* Debugger communication via TCP/IP connection.

connection is only the update information on processor resource values for the debugger frontend as well as control information on pushed buttons in the frontend that need to be processed by the simulator. However, in most cases the user sets breakpoints in the application code and runs the simulator till the breakpoint is hit. During that time, the graphical debugger frontend needs no update. It is obvious that doing the complete communication between the system and the software simulator via function calls without the need to update the frontend increases overall simulation speed significantly. The simulation speeds achieved using this very flexible mechanism are close to the simulation without debugger frontend (cf. section 3.1). The reason for not achieving the same results are founded in the need to frequently poll for occurring events in the graphical debugger frontend. The events could e.g. come from someone pushing the *stop* button during simulation run.

## 3.3 Multiple Processor Cores

Today, complex platforms are often composed of a large variety of programmable components. It is evident that besides system simulation speed another hurdle is brought up when debugging such systems: managing all the debugger frontends. The problem of administrating several debuggers is depicted in figure 8.5.

Indeed, not all debugger frontends are required at the same time but it is desirable to be able to set a breakpoint in a specific processor, run the system to a point where the breakpoint is hit and then continue debugging the respective processor instruction by instruction in the debugger frontend.

The graphical debugger provided by the system integrator of the LISA processor design platform is capable of administrating multiple processor cores

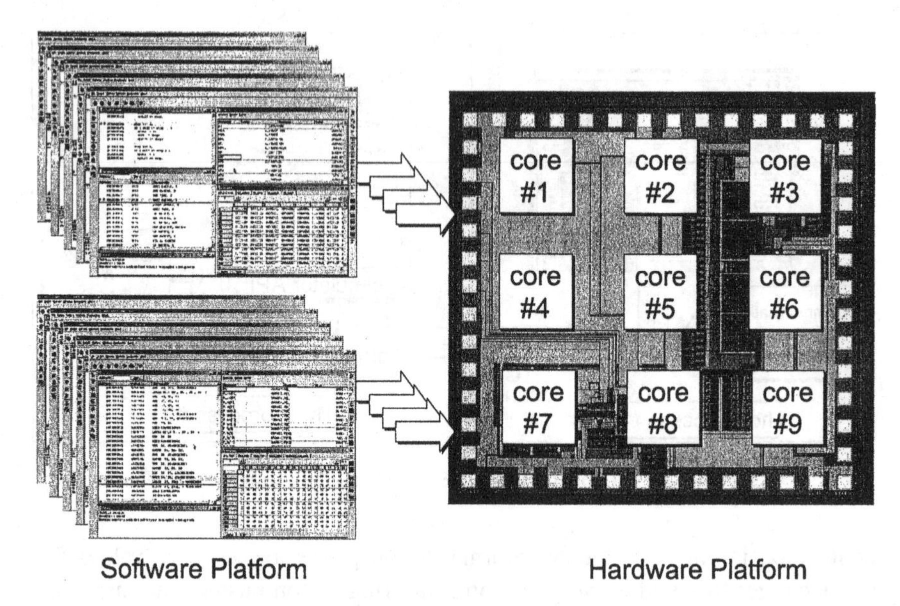


*Figure 8.5.* Multiple debuggers in a system comprising a number of processor cores.

within one single debugger frontend. At the time of instantiation, the user determines which processor core to be administrated by which debugger frontend. In contrast to the graphical debuggers presented in chapter 7.3, this special system integration debugger can switch dynamically between the core that needs special attention during run-time. Figure 8.6 shows the same system as in figure 8.5, however, this time, all cores are administrated by one single debugger.

Only one core is visible at a time in the multi-processor debugger frontend. Thus, the TCP/IP connections between the debugger frontend and the other cores administrated by the debugger frontend are not active. When switching from one core to the next, the connection is established and the debugger is configured to the processor under investigation. In case one of the administrated processor cores hits a breakpoint in run-mode, it is automatically focused in the debugger and the complete system simulation is paused.

Using this flexible debug mechanism of multiple cores within few graphical debuggers, full visibility of all cores is combined with clearness and high system simulation speed, as only those cores communicate via TCP/IP connection that are also visible in the debugger.

## 4. Case Study: CoCentric System Studio

In a case study in cooperation with Synopsys Inc., Mountainview, USA, LISA processor simulators were integrated into their commercially available

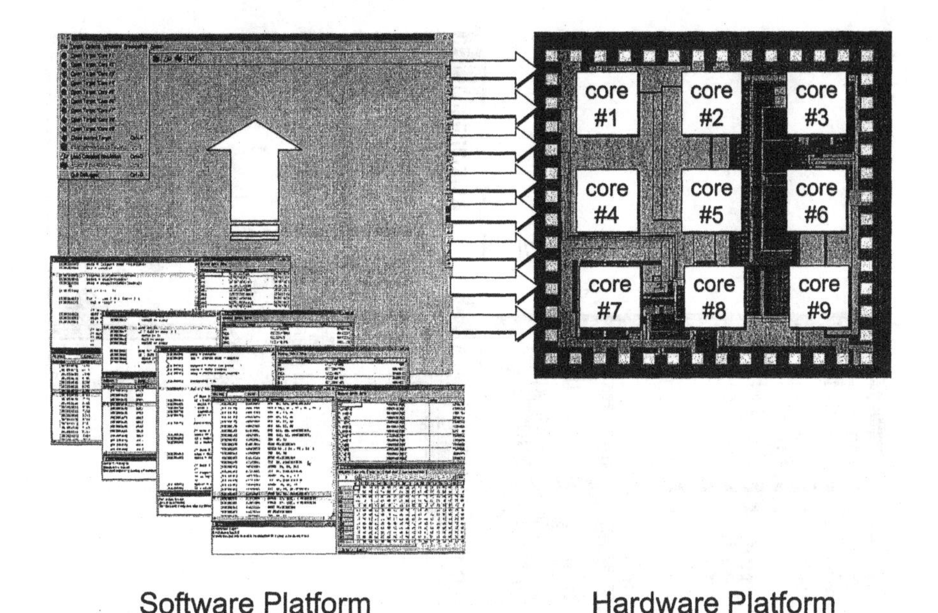


*Figure 8.6.* Single debugger administrating a number of processor cores.

tool *CoCentric System Studio* (CCSS) [26]. CCSS is a system-level design environment for the rapid creation of executable system specifications that can be verified and implemented as hardware and software functions. It enables designers to use multiple hierarchical, graphical and language abstractions in order to capture system complexity in one unified environment based on C, C++ and SystemC.

LISA processor simulators are wrapped with an additional, CCSS-specific API, which allows seamless integration into the system. Figure 8.7 shows a small system set up in CCSS consisting of two processors which are accessing external memories via a joint bus, a bus arbiter, and an external clock pushing forward simulation.

The LISA processor simulators can be chosen from a library within CCSS and included into the system. In the graphical view of the system, the processor models are depicted as large square boxes. The ports of these boxes are the system clock, which pushes the simulators forward, and the pins of the processor as specified in the underlying LISA model. Once the system simulation is initialized, the debugger frontend(s) are launched. In the sample system shown in figure 8.7, both processors are administrated by one debugger (cf. section 3.3). The simulation can be controlled both via CCSS and the graphical LISA debugger frontend.

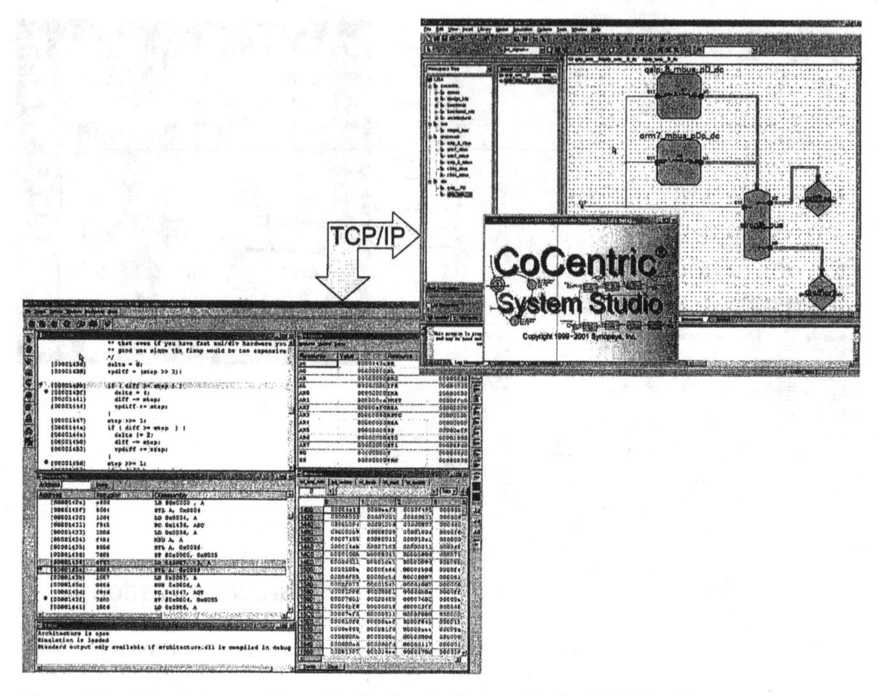


*Figure 8.7.* Simulator coupling with CoCentric System Studio (CCSS) of Synopsys.

As explained in section 3.2, the system simulation runs at nearly the same speed as it would without the graphical debugger frontends in case the processor simulators are in run-mode. Once a breakpoint is hit in either the software simulators or the system[1], the simulation stops. In case the breakpoint occurred in the software simulators, the respective processor is focused on in the graphical debugger frontend.

## 5. Concluding Remarks

In this chapter the fourth quadrant of the LISA processor design platform was introduced – system integration and verification. Motivated by the ever increasing system complexity and by the upcoming of platform-based designs integrating heterogeneous components, it was shown that new system design methods are required. To cope with these requirements, a framework for *virtual prototyping* was introduced which allows integration of heterogeneous components on various abstraction levels.

[1]CoCentric System Studio allows the setting of breakpoints in the underlying C++/SystemC code.

On the software side it was shown how LISA processor simulators can be integrated into such a prototyping environment on various levels of abstraction of their underlying processor model. Furthermore, it was pointed out that system simulation speed is essential to cope with the rising number of test-vectors required for verification. For this reason, the simulators generated from LISA processor models use a TCP/IP based mechanism to communicate with the graphical debugger frontend. This mechanism combines full visibility of the simulation with highest simulation speed. Besides, the debugger used in the system integration phase is capable of administrating multiple cores. This increases clearness during simulation while keeping full visibility of all cores in the system and maximizing system simulation speed. Finally, the seamless integration of LISA processor simulators into CoCentric System Studio was shown.

# Chapter 9

# SUMMARY AND OUTLOOK

Within the scope of this book a methodology supporting the efficient design of ASIPs has been presented: the LISA processor design platform (LPDP). The design of ASIPs is a long, tedious, and error-prone task consisting of four design phases – architecture exploration, architecture implementation, application software design, and system integration and verification. Without automation and a unified environment, the future design of such ASIPs is limited to very few semiconductor companies which have expertise in each of the above mentioned fields.

The LPDP approach overcomes this hurdle as the complete processor design process is based on one sole description of the target architecture in the LISA language. In the architecture exploration phase, the processor model in LISA can be stepwise refined from the most abstract level to the level of micro-architectural implementation. On every level of accuracy, a working set of software development tools can be generated automatically comprising code generation tools (assembler, linker) and simulation/evaluation tools (simulator, debugger, profiler). The profiling capabilities of the software simulator are used to gather execution statistics of the application very early in the exploration process, measure the usage of instructions once the instruction-set is introduced to the model, and determine the frequency of accesses to processor resources like registers, functional units, and pipeline stages on the micro-architectural implementation level.

Once the micro-architecture is fixed, the architecture is implemented using the HDL code generator of the LPDP environment. Here, large parts of the synthesizable implementation model in a HDL like VHDL and Verilog can be generated. The generated parts of the architecture comprise those parts that are long, tedious, and error-prone to write manually, but have less influence on the overall performance: structure, decoder, and controller. Using an image, these parts of the architecture could be referred to as the *brain* whereas the data-path, which represents the crucial portion of the architecture in terms of critical path, power consumption, and speed, could be referred to as the *muscles*. The latter carries the designers intellectual property (IP) and thus needs to be specified and inserted manually into the generated model. The proof of concept was carried out on a case study of a DVB-T ASIP designed in cooperation with Infineon Technologies.

To enable application designers to comfortably program the architecture, a set of production quality software development tools is needed that is comparable in speed and functionality with tools provided with commercially available processor architectures. It was shown in four case studies that the tools derived from LISA processor models can compete well both in functionality and speed with different vendor tools. As the generated simulation tools are enhanced in speed by the compiled simulation technology they are faster by one to two orders in magnitude than the simulators of the processor vendors (on the same level of accuracy), which run exclusively in interpretive mode.

Finally, the LPDP environment derives flexible co-simulation interfaces for seamless simulator integration into arbitrary system simulation environments. In a case study, the generated processor simulators were integrated into the commercial CoCentric System Studio environment of Synopsys. Moreover, to cope with the increasing number of processor cores on a single chip, the observability of such systems was enhanced by allowing the monitoring of several processors in the system in one sole debugger frontend.

To summarize, it was shown within the scope of this book that the challenges and problems associated with the design of ASIPs are addressed and can be overcome by the LISA-based processor design approach. In the following sections, further research topics that are associated with the LISA-based processor design and require further investigation are introduced. This concerns processor modeling capabilities in general as e.g. memory modeling, and enhancements of the LPDP approach in each of the presented processor design phases.

## 1. Processor Modeling

Although LISA allows modeling of a wide range of processors with architectural characteristics like VLIW, SIMD, MIMD, superscalarity, etc. there are still some limitations in modeling particular aspects of processor architectures (cf. chapter 3.3.3). Besides these, there is currently no formalism in the language to model peripherals such as timers, interrupt controllers, and PIOs (parallel input-output). However, as LISA is based on C/C++ it is feasible to build a library of functions mirroring the functionality of the respective components which can later be compiled with a native C/C++ compiler of the host system and linked to the executable simulator.

On the other hand and more importantly, memory modeling is limited to simple array structures in the current version of the LISA language (cf. chapter 3.2.1). For this purpose, the language has to be enriched with a formalism to describe memory subsystems of arbitrary structure.

## 1.1 Memory Hierarchies

System-on-chip (SOC) technology permits customization of processor cores with different memory architectures. For this reason, joint system and architecture design also requires the exploration of different memory hierarchies and organizations. At the current stage of research, the LISA language contains no formalism to describe arbitrary memory structures. As with the description of peripherals, it is feasible to provide the memory model in form of a C/C++ library. Latencies between the request and the availability of data from memory can be taken into consideration e.g. by manually stalling the pipeline. Obviously, this manual approach is tedious and error-prone and prohibits fast exploration of different memory configurations, like e.g. cache hierarchies and strategies. As especially caches are playing an increasingly important role in the area of DSP and ASIP architectures, future work on LISA should research common memory systems and enrich the language by a qualified formalism for their description.

Besides the extension of the LISA language towards memory modeling, the simplification of the transition from instruction- to cycle-accuracy is worth further investigation. This is briefly discussed in the following.

## 1.2 Seamless Processor Model Refinement

In the architecture exploration phase, the processor model of the target architecture is refined from an abstract, application-centric level to the level of register-transfer accuracy. While the gap between the different levels of abstraction in the first refinement steps is relatively small, the opposite is true for the step from the instruction-set accuracy to cycle-accuracy. This step is characterized by the introduction of the micro-architecture to the model comprising architectural details like pipelining and the respective distributed execution of instructions over several pipeline stages.

Typically, the code generation tools are derived from the instruction-based model as it contains the required information on the instruction-set and the abstracted behavior of the target architecture to retarget high-level language (HLL) C-compiler, assembler, and linker. Once the instruction-set is fixed, several micro-architectures are tried out that implement the respective instruction-set. However, in the current version of the LISA language, every time this step is carried out a new model has to be realized that contains both the instruction-set model and the micro-architectural implementation. This step is error-prone, as the instruction-set has to be specified twice – in the instruction-based model that retargets the code generation tools and in the cycle-based model that retargets the simulator.

A solution to this problem is a clear separation between instruction-set and abstracted behavior on the one hand and structure and detailed behavior on the other hand. This could be achieved, e.g. by adopting a construct which is well known from object oriented programming languages like C++ to the LISA language: *inheritance*. Inheritance could be applied to LISA operations by assigning the instruction-set and abstracted behavior information to a *base* operation which operations in the cycle-based model are derived from. By this, they automatically inherit the properties not specified in the derived operation.

## 2. Architecture Exploration

At a certain point in the processor design process, the instruction-set is introduced to the model. As shown in chapter 5, this is the case in the refinement step from the data-path to the instruction-based model. Especially the specification of the binary instruction word coding is a difficult task as here conflicting design constraints need to be considered: orthogonality for HLL-compiler friendliness versus instruction-word compression and non-orthogonality for low power consumption.

At the current stage of research, the binary coding has to be specified and optimized manually. This concerns the fixing of the instruction-word width and the placement of opcode, register, and immediate fields within the instruction-word. Future work should concentrate on developing methodologies to automatically generate the optimal binary coding of instructions given a set of constraints with their weighting.

## 3. Software Development Tools

Originally, the LISA language was developed to retarget fast compiled processor simulators with dynamic scheduling [97]. In recent years, it has been extended to also address the generation of further software development tools like assembler, linker, and graphical debugger frontend. Besides, dedicated language elements were added to support the generation of synthesizable HDL code and system integration to address the complete processor design flow.

However, to efficiently explore and program the target architecture it is required to abstract from machine dependent assembly code to HLL code (cf. chapter 7.1). For this reason, the LISA language needs to be extended to enable the retargeting of HLL C-compilers. This extension includes the ability to specify constraints and calling conventions.

Work on compiler retargeting using LISA has already been started. In order to overcome the entrance barriers related to compiler design like C-code parsing and the specification of an appropriate intermediate format, an existing compiler framework has been chosen. The CoSy system [76] of Associated Compiler Experts bv. (ACE), Amsterdam, The Netherlands, provides a compiler infrastructure including parser, intermediate representation (IR), and interface functions to access the IR that allows to concentrate on architecture related issues.

## 4. Architecture Implementation

At the current stage of research, synthesizable HDL code for control-path, structure, and instruction decoder is generated from LISA processor models in the architecture implementation phase. Besides, wrappers to functional units and the interconnects to processor resources, such as registers and memories, are created. However, the content of the functional unit has to be added manually in an HDL like VHDL or Verilog. Obviously, this introduces consistency problems between the hardware implementation model and the underlying LISA model. Momentarily, this problem is addressed by automatically generated test-programs and stimuli [190], which are used to verify the software model against the HDL model by simulation.

With the advent of the SystemC modeling language [183], this hurdle can be overcome. The SystemC language combines the syntax of the C++ programming language with the semantics of a hardware description language. Used at register-transfer (RT) accuracy, SystemC code is fully synthesizable [186].

### 4.1 Data-Path Synthesis

According to chapter 3.2.3, the behavioral model specified in LISA is based on pure C/C++ code. At the current stage of research, this behavioral code is discarded during the architecture implementation phase and rewritten manually in the target HDL.

However, as SystemC represents a subset of the C++ language, the C++-code used in the behavioral model of the LISA description can on the one hand be simulated in the software simulator and on the other hand taken into consideration during architecture synthesis [191]. It is clear, that the behavioral C++-code has to be restricted to those parts, that are synthesizable in SystemC [186]. This limitation concerns e.g. the usage of pointers.

It was argued in chapter 6 that the data-path in an architecture is crucial in several respects and is thus often realized using full custom design. The proposed data-path generation using SystemC does not contradict this as it is the goal to get a first working *reference* HDL model which the hand-optimized version can be verified against.

In the context of the data-path generation, resource sharing is a topic that requires attention as it has a strong impact on the size and power consumption of the hardware implementation model. Extraction of information indicating resource sharing is discussed in the following.

## 4.2 Resource Sharing

In case of generating HDL code for the data-path of the target architecture from the LISA model, the obvious fact is neglected that not all operations within functional units are active at once. A manual realization would share resources within one unit to save chip area and power consumption.

LISA allows the specification of functional units in the model, which operations can be assigned to (cf. chapter 3.2.6). Using this information combined with an automated analysis of the model for possibly concurrently active operations, the required information to share resources could be extracted and taken into consideration for data-path generation.

# 5. Concluding Remarks

It is obvious, that the above mentioned aspects just represent a small subset of potential research areas related to the LISA processor design platform. Due to the width of this subject which is worked on by both software tool designers, computer architects, and system designers the area of interesting research topics is endless.

# Appendix A
# Abbreviations

| | |
|---|---|
| ACE | Associated Compiler Experts bv. |
| ADPCM | Adaptive Differential Pulse Code Modulation |
| ALU | Arithmetic Logical Unit |
| API | Application Programming Interface |
| ASCII | American Standard Code for Information Interchange |
| ASIC | Application-Specific Integrated Circuit |
| ASIP | Application-Specific Instruction-Set Processor |
| ASSP | Application-Specific Signal Processor |
| ATM | Asynchronous Transfer Mode |
| CCSS | CoCentric System Studio |
| COFF | Common Object File Format |
| CORDIC | Cordinate Rotation Digital Computer |
| CPU | Central Processing Unit |
| DECT | Digital Enhanced Cordless Telecommunications |
| DSP | Digital Signal Processor |
| DVB-T | Digital Video Broadcast Terrestrial |
| EDA | Electronic Design Automation |
| EP | Embedded Processor |
| ETSI | European Telecommunication Standardization Institute |
| FFT | Fast Fourier Transformation |
| FPGA | Field Programmable Gate Array |
| GRACE | Generic ATM Cell Stream Processor |
| GSM | Global System for Mobile Communications |
| HDL | Hardware Description Language |
| HDTV | High Density Television |
| HLL | High Level Language |

| | |
|---|---|
| IC | Integrated Circuit |
| IP | Intellectual Property |
| IPC | Inter-Process Communication |
| IR | Intermediate Representation |
| ISA | Instruction-Set Architecture |
| ILP | Instruction-Level Parallelism |
| ITU | International Telecommunication Union |
| LISA | Language for Instruction-Set Architecture |
| LOF | LISA Object File Format |
| LPDP | LISA Processor Design Platform |
| $\mu$C | Micro-Controller |
| MIMD | Multiple Instruction Multiple Data |
| MIPS | Million Instructions per Second |
| MOPS | Million Operations per Second |
| MSB | Most Significant Bit |
| OTS | Off-the-Shelf |
| PDA | Personal Digital Assistant |
| PIO | Parallel Input Output |
| PPU | Post Processing Unit |
| QFC | Quantum Flow Control |
| QoS | Quality of Service |
| RAM | Random Access Memory |
| RISC | Reduced Instruction-Set Computer |
| RTL | Register Transfer Level |
| SIMD | Single Instruction Multiple Data |
| SLI | System Level Integration |
| SOC | System-on-Chip |
| SPC | Section Program Counter |
| SPEC | Standard Performance Evaluation Corporation |
| VHDL | Very High Speed Integrated Circuit HDL |
| VLIW | Very Large Instruction Word |
| VLSI | Very Large Scale Integration |

# Appendix B
# Grammar of the LISA Language

This chapter presents the grammar of the LISA language. It shall be mentioned though that the grammar is incomplete since the complete grammar of the LISA language would go beyond the scope of this document. As already indicated in the preface of this book, the language and the tooling presented are only a snap-shot of the research work carried out at the Institute for Integrated Signal Processing Systems (ISS) at Aachen University of Technology (RWTH Aachen). To obtain the latest information please refer to the webpages of the ISS at "http://www.iss.rwth-aachen.de/lisa" or for commercial use at LISATek at "http://www.lisatek.com".

## Fundamental Definitions

To simplify the understanding of the following sections and to shorten the description, some fundamental definitions of identifiers are made in this section. The definitions of these designators are used within the rest of this section.

**Fonts:** Within the remaining section, different fonts distinguish several elements. Terminal elements are illustrated in a `Typewriter` font. Non-terminal elements are visualized in an *Italic* font.

**Keywords:** In table B.1, all LISA keywords are listed. Within the remaining section, these keywords are illustrated with a sanserif font.

*Table B.1.* LISA language keywords.

| List of LISA language keywords | | |
|---|---|---|
| ACTIVATION | ADDRTYPE | ALIAS |
| ALWAYS | AT | BANKS |
| BEHAVIOR | BLOCKSIZE | BUS |
| bit | BYTES | CACHE |
| CASE | CODING | CONTROL_REGISTER |
| CURRENT_ADDRESS | DECLARE | EXPRESSION |
| ENTRY_ADDRESS | ENUM | FLAGS |
| GROUP | IF THEN ELSE | IN |
| INSTANCE | LABEL | LINESIZE |
| LATENCY | main | MEMORY_MAP |
| MEMORY | NEVER | NUM_WAYS |
| OPERATION | PAGE | PIN |
| PIPELINE | PIPELINE_REGISTER | PROGRAM_COUNTER |
| RAM | RANGE | REGISTER |
| reset | RESOURCE | REFERENCE |
| RPL_POLICY | SIZE | SUBOPERATION |
| SWITCH | SYMBOL | SYNTAX |
| TClocked | USERTYPE | WA_POLICY |
| WB_POLICY | | |

## Fix designators

In the following paragraphs fix designators are defined, which are being used without further explanation. In addition to the syntax, the meaning of the respective designators is given.

*instance-name:*
 *operation-name*

$member_1$ .. $member_n$:
 *operation-name*

*reference-name:*
 *group-name*

$group\text{-}name_1$ .. $group\text{-}name_n$:
 *identifier*

*operation-name:*
    *identifier*

*suboperation-name:*
    *identifier*

*label-name*$_1$ .. *label-name*$_n$*:*
    *identifier*

*pipeline-name:*
    *identifier*

*element-name:*
    *identifier*

*stage*$_1$ .. *stage*$_n$*:*
    *identifier*

*identifier:*
    *upper-case-letters* { *upper-case-letters* | *lower-case-letters* | *number* | *underbar* }
    | *lower-case-letters* { *upper-case-letters* | *lower-case-letters* | *number* | *underbar* }

*upper-case-letters:*
    `A | B | C | D | E | F | G | H | I | J | K | L | M | N |`
    `O | P | Q | R | S | T | U | V | W | X | Y | Z`

*lower-case-letters:*
    `a | b | c | d | e | f | g | h | i | j | k | l | m | n |`
    `o | p | q | r | s | t | u | v | w | x | y | z`

*number:*
    *decimal-type-number*
    | *binary-number*

*decimal-type-number:*
    *decimal-number*
    | *hexadecimal-number*

*binary-number:*
    *binary-number-prefix binary-number-element* { *binary-number-element* }

*binary-number-prefix:*
    `0b`

*binary-number-element:*
`0 | 1`

*decimal-number:*
*decimal-number-element* { *decimal-number-element* }

If a number is used without any prefix like `0x` or `0b` it is interpreted as a decimal number.

*decimal-number-element:*
`0 | 1 | 2 | 3 | 4 | 5 | 6 | 7 | 8 | 9`

*hexadecimal-number:*
*hexadecimal-number-prefix hexadecimal-number-element*
{ *hexadecimal-number-element* }

*hexadecimal-number-prefix:*
`0x`

*hexadecimal-number-element:*
`0 | 1 | 2 | 3 | 4 | 5 | 6 | 7 | 8 | 9 | a | b | c | d | e | f`

*underbar:*
`'_'`

*lisa-array:*
`'['` *lower-array-boundary* `..` *upper-array-boundary* `']'`

*lower-array-boundary:*
*decimal-number*

*upper-array-boundary:*
*decimal-number*

*bit-size:*
*decimal-number*

*bit-width:*
*decimal-number*

*address:*
*hexadecimal-number*

*lisa-data-type:*
  *C-data-types*
  | *C++-data-types*
  | *user-defined-types*
  | *generic-integer-type*
  | *TClocked-type*

*cond-expression:*
  *CStatement*

*expression-type:*
  *C-data-types*
  | *user-defined-types*

*CStatement:*
A CStatement may comprise any possible sequence of statements that is written in the programming language C.

*C-data-types:*
Native data types of the programming language C.

*C++-data-types:*
Native data types of the programming language C++.

*user-defined-types:*
Data types which are defined by the user and based on C/C++ native data types.

*generic-integer-type:*
  *type-specifier* `bit` '[' *bit-width* ']'

*type-specifier:*
  `signed` | `unsigned`

*TClocked-type:*
Data type which is used to describe cycle-true behavior of resources. In cycle-accurate models, resources labelled with a *resource-specifier* are simulated with a cycle-true behavior.

*lisa-comment:*
    *single-line-comment* | *multi-line-comment*

*single-line-comment:*
    *slc-specifier single-line-comment-text*

*multi-line-comment:*
    *mlc-begin-specifier multi-line-comment-text mlc-end-specifier*

*slc-specifier:*
    //

*mlc-begin-specifier:*
    /*

*mlc-end-specifier:*
    */

# LISA Architecture Model

A LISA architecture model consists of two main parts. The first part comprises the description of processor resources which takes place in the resource-section of the model. The second part is embodied in so-called LISA operations. They model the behavior and the instruction-set of the target architecture.

*LISA description:*
    *ResourceSection* { *ResourceSection* }
    *LISA Operation* { *LISA Operation* }

# The Resource-Section

Within the resource-section of a LISA model all processor resources such as pipelines, memories, registers, etc. are defined. Thus, the fundamental structure of the target architecture is given by this section. The RESOURCE section consists of several components: a *MemoryMap*, *ResourceElements*, and the definition of one or more pipelines with their respective pipeline registers. Within the *MemoryMap* the virtual address space of the architecture is mapped onto processor resources, and the bus/memory configuration is described. The syntax of the resource section is visualized in the following.

*ResourceSection:*
RESOURCE '{'
[ *MemoryMap*$_1$ ... { *MemoryMap*$_n$ } ]

[ *MemoryElement*$_1$ ... { *MemoryElement*$_n$ } ]

[ *ResourceElement*$_1$ ... { *ResourceElement*$_n$ } ]

[ *AliasStatement*$_1$ ... { *AliasStatment*$_n$ } ]

[ *Pipeline*$_1$ [ *PipeRegister* ] ... { *Pipeline*$_n$ [ *PipeRegister* ] } ]
'}'

*MemoryMap:*
MEMORY_MAP [ *map-ident* ] '{'
*AddressSpec*$_1$ -> *ResourceSpec*$_1$ ;
...
*AddressSpec*$_n$ -> *ResourceSpec*$_n$ ;
'}'

*AddressSpec:*
[ BUS '(' *bus-ident* ')' ',' ]
RANGE '(' *lower-address* ',' *upper-address* ')'
[ ',' PAGE '(' *page-number* ')' ]

*ResourceSpec:*
*memory-ident* [ '[(' *upper-bank-bit* '..' *lower-bank-bit* ')]' ]
'[(' *upper-addr-bit* '..' *lower-addr-bit* ')]'

The memory map provides information about the link between virtual address space and physical memory, for both initializing the memory and accessing it via optional buses. The RANGE keyword in the memory map defines a portion of the virtual address space to be mapped onto an existing physical resource. This resource is specified in the *ResourceSpec* and must be present in the model, i.e. defined in the resource-section. How addresses are mapped onto a resource is defined by the bit specifications in parenthesis in the *ResourceSpec*. These bit ranges specify the portion of the address which is used as an index for the respective resource. Optionally, a bank address bit range might be given to select the part of the address word that is used for bank selection. A memory map can contain an arbitrary number of such mappings. Each of these lines consists of an *AddresseSpec* and a *ResourceSpec* and is terminated with a semicolon. The syntax for the actual definition of the memories and buses is provided below.

*MemoryElement:*
*memory-type memory-data-type memory-ident* '{'
*parameter-ident*$_1$ '(' *parameter-value* ');'
...
*parameter-ident*$_n$ '(' *parameter-value* ');'
'}'

*memory-type:*
```
    CACHE
    | RAM
    | MEMORY
    | BUS
```

*parameter-ident:*
```
    SIZE
    | NUM_WAYS
    | LINESIZE
    | ADDRTYPE
    | BLOCKSIZE
    | ENDIANESS
    | WA_POLICY
    | WB_POLICY
    | RPL_POLICY
    | LATENCIES
    | CONNECT
    | USERTYPE
    | FLAGS
    | WRITEBUFSIZE
```

Four different types of memories are supported. It is recommended to use CACHE, BUS, and RAM in favor of MEMORY, since the latter only defines a simple C-array, which does not support access via the memory API functions. However, for abstract models in which latencies and cycle-accuracy are not of major importance, these simple memory implementations can be used to achieve highest simulation performance.

The syntax for other resources like registers or pipelines is given in the following.

*ResourceElement:*
    [ *ResourceSpecifier* ] *lisa-data-type element-name* [ *lisa-array* ] ';'

*ResourceSpecifier:*
```
    CONTROL_REGISTER
    | PIPELINE_REGISTER
    | PROGRAM_COUNTER
    | PIN
    | REGISTER
```

**ResourceElement:** Resources within a LISA model can be used with or without a *resource specifier*. E.g. the program counter is labelled with the LISA *resource specifier* PROGRAM_COUNTER. One resource element is defined by its data type *lisa-data-type* and the name of the resource *element-name*. If it is an array type the array has to be defined additionally.

**Alias:** Frequently, DSPs feature registers which are combined of multiple bit-fields (respectively sub-registers). These bit-fields can be accessed independently as well as a whole. Furthermore, many processor architectures provide so-called memory-mapped resources, like e.g. registers of peripherals. This eases software development without affecting performance or chipsize of the processor, and has therefore become a popular feature not only in DSP architectures.

Both architectural properties can be elegantly modelled using the LISA concept of *alias resources.*

*AliasStatement:*
*alias-data-type alias-element* '{' ALIAS '{' *corresp-element* ';' '}' '}'

*alias-element:*
*alias-name* [ *lisa-array* ]

*corresp-element:*
*corresp-resource-element* [ '[' *start* ']' ]

*alias-name:*
*identifier*

*corresp-resource-element:*
*ResourceElement*

It is important to understand that aliases do not define new states. They rather define pointers to resources that are defined in the model.

The declaration of aliases to resources follows the declaration style of simple resources. The alias data-type must be the same as the data type of the referenced resource. The declaration is introduced by the keyword ALIAS followed by curly braces enclosing the name of the resource the aliasing refers to (*corresp-element*). If the alias refers to a position in an array of resources, the index of the resource array that the aliased resource points to (*start*), needs to be specified in brackets.

**Pipeline, pipeline register:** The syntax for the definition of a pipeline and its pipeline registers is shown below.

*Pipeline:*
PIPELINE *pipeline-name* = '{' $stage_1$ { ';' $stage_n$ } '}' ';'

*PipeRegister:*
PIPELINE_REGISTER IN *pipeline-name* '{'
$PipeRegElement_1$
...
{ $PipeRegElement_n$ }
'}' ';'

*PipeRegElement:*
*lisa-data-type element-name* [ *lisa-array* ] ';'

A pipeline definition in LISA starts with the keyword PIPELINE followed by a unique name *pipeline-name* and the enumeration of its stages separated by semicolons and enclosed in curly braces. The ordering corresponds to their spatial ordering in the processor hardware. The pipeline registers between each consecutive stage are defined separately by the *resource specifier* PIPELINE_REGISTER.

Pipeline registers are always defined for a single pipeline. To do so, the *resource specifier* PIPELINE_REGISTER is used, followed by the keyword IN, the *pipeline-name* and the pipeline register elements. A pipeline register element consists of the data type, the name of the element and possibly the array definition. Such a declaration is terminated by a semicolon.

**Pipeline Control Functions:** As mentioned above, the pipeline registers are defined in the resource-section of the LISA model of an architecture. By definition, all pipeline registers are multiple instantiated and exist between each consecutive pipeline stage.

Both, the complete pipeline and a single pipeline register may be controlled via a set of pipeline control functions which are part of the generic pipeline model underlying the LISA language. LISA provides four different pipeline control functions:

- execute,
- shift,
- stall and
- flush.

The execution of the state-update function is performed explicitly by using the predefined function `execute`, which is called as follows:

`PIPELINE(` *pipeline-name* `).execute()`

Upon calling this function, the respective behavior of the currently processed instructions is executed. The execution ordering of the stages is from the last to the first stage. The pipeline function `shift()` is called according to the function `execute()`. Moving the values of the pipeline registers from the input to the output, this pipeline function simulates the clock edge of the real hardware. The function `stall()` can be used for the whole pipeline or for a single pipeline register. The pipeline function `stalled()` can be used to check whether a pipeline register is currently stalled. If the respective stage is stalled, a non-zero value is returned. The pipeline function `flush()` deletes the values in the specified pipeline register. For this reason, all pipeline register elements are reset to zero.

These above mentioned control functions are called by

`PIPELINE_REGISTER(` *pipeline-name* `).shift()`
`PIPELINE_REGISTER(` *pipeline-name* `).stall()`
`PIPELINE_REGISTER(` *pipeline-name, stage1/stage2* `).stalled()`
`PIPELINE_REGISTER(` *pipeline-name, stage1/stage2* `).stall()`
`PIPELINE_REGISTER(` *pipeline-name, stage1/stage2* `).flush()`

## LISA Operation

In LISA, two different kinds of operations exist – *alias-operations* and *ordinary-operations*. In general, two instructions within one LISA model must not have the same coding. However, the assembly language of many processors comprises different instructions which are mapped onto the same instruction word, since one instruction is a specialty of the other. For this reason the `ALIAS` keyword exist in the LISA language. The syntax of a LISA operation is visualized in the following.

*LISA Operation:*
*alias-operation | ordinary-operation*

An *alias-operation* is structured as an *ordinary-operation*, however, it has the same coding as another LISA operation.

*alias-operation:*
ALIAS OPERATION *operation-name* [ IN *pipeline-name.stage* ] '{'
[ *DeclareSection* ]
[ *CodingSection* ]
[ *SyntaxSection* ]
[ *BehaviorSection* ]
[ *ActivationSection* ]
[ *ExpressionSection* ]
'}'

As mentioned above a LISA OPERATION consists of several sections to describe parts of the instruction-set as e.g. the coding and the syntax. One or more operations form an instruction.

Frequently, similar operations occur which share the same coding (e.g. different register files or processing modes) but differ in their syntax or behavior description. This is formalized in LISA by using so-called sub-operations within regular operations. Sub-operations allow to create several operations which are grouped under the same main operation identifier. These operations have their own sub-names to permit access. They are introduced by the keyword SUBOPERATION followed by an identifying name and curly braces enclosing the section definitions *coding*, *syntax*, *expression*, *behavior*, and *activation*.

*ordinary-operation:*
OPERATION *operation-name* [ IN *pipeline-name.stage* ] '{'
*operation-body* { *operation-body* }
'}'

A LISA operation is introduced by the keyword OPERATION followed by an identifying *operation-name*. If a pipelined architecture is modelled cylce-accurately, the operation has be assigned to a certain pipeline stage of the processor. This is done using the keyword IN, the *pipeline-name* and, separated by a dot, the *stage* of the pipeline in which the operation is performed. Subsequently, enclosed in curly braces, the declaration of sub-operations or the different sections of the LISA operation occurs.

*operation-body:*
SUBOPERATION *suboperation-name section-list*
| *section-list*

*section-list:*
[ *DeclareSection* ]
[ *CodingRootSection* ]
[ *CodingSection* ]
[ *SyntaxSection* ]
[ *BehaviorSection* ]
[ *ActivationSection* ]
[ *ExpressionSection* ]

In contrast to ordinary operation groups, the selection of the sub-operation is not performed based on the coding (which is identical for all sub-operations) but expressed explicitly by the reference which is composed of the names of the main operation and the respective sub-operation separated by a dot.

*operation-name.suboperation-name*

## Declare-Section

The declare-section is the first section to be specified within a LISA operation. In this section, local identifies used within the operation as well as alternative non-terminal elements are declared.

*DeclareSection:*
  DECLARE '{'
  *Item_1*
  ...
  { *Item_n* }
  '}'

*Item:*
  *Instance*
  | *Group*
  | *Reference*
  | *Label*
  | *Enum*

A declare section is introduced with the keyword DECLARE, followed by the declarations which are embraced by curly braces. The different items within this section can be *Instances* of operations, *Groups* of operations, *References* to other operations, *Labels* and *Enums*. These items are introduced in the following sections in more detail.

**Instance:** An *instance* defines a reference to a specific operation and is declared in the declare section of a LISA operation by prefixing the operation name with the keyword INSTANCE. If an instance is used within a coding section, each element stands for the coding of the corresponding operation. Therefore it is implied, that the referenced operation contains a coding section as well. The same can be applied to syntax and expression section. Within the behavior section, the instance embodies the expression section of the respective operation. If it is made use of an instance within the activation section, it always refers to activation *and* behavior sections. If more than one operation reference occurs, the identifiers following the keyword INSTANCE are separated by a comma.

*Instance:*
  INSTANCE *operation-name*$_1$ { ',' *operation-name*$_n$ } ';'

**Group:** References to a list of alternative operations are formalized using the GROUP keyword in LISA. If such a group name is used within any section of a LISA operation it is expressed, that several alternatives are allowed at this point. It is mandatory that the corresponding operations also contain a section of the kind of the section the group is used in. The syntax of a group definition is presented in the following. The names of the group members are the ones of several LISA operations.

*Group:*
  GROUP *group-name*$_1$ { ',' *group-name*$_n$ } = '{'
  *member*$_1$ { '||' *member*$_m$ } '}' ';'

Furthermore, groups may be accessed from other operations via the reference mechanism, which is introduced in the next paragraph.

**Reference:** To cope with the hierarchical structuring of LISA operations, it is frequently required to reference groups, which are not defined in the current operation. LISA provides a referencing mechanism, which enables the designer to access elements of groups, which are defined in superior operations. Superior means, that the operations defining these groups are specified in an upper level of the operation hierarchy as the operation requesting access to these groups. Besides the coding-section, it is allowed to use referenced groups in all sections within operations and also within the LISA control-structures.

*Reference:*
REFERENCE *group-name*$_1$ { ',' *group-name*$_n$ } ';'

**Label:** A label might be understood as a local variable (that means: local to the current operation), which can be used to establish a link between the different sections within an operation. A value can be assigned to this label in the coding and/or syntax section, and the value of the label can be used in any other section. A label is a symbolic identifier for a numeric value. Labels must be declared in the declare section of the respective LISA operation. The syntax definition is given below.

*Label:*
LABEL *label-name*$_1$ { ',' *label-name*$_n$ } ';'

## Coding-Root Section

Instructions and their operands are recognized in the processor hardware by gradually decoding the logical bit fields of the instruction word that is held in the instruction register. The corresponding structure in LISA descriptions begins in the *coding root* operation. In order to decode instructions, the coding patterns defined by the totality of all described operations must be compared to the actual value of the current instruction word (or even multiple instruction words). Therefore, the resource holding the instruction word for decoding and the address the instruction is fetched from is specified in the coding root operation.

*CodingRootSection:*
*cond-coding-root* | *non-cond-coding-root*

In general two main kinds of coding roots exist: conditional coding roots and the non conditional coding roots. While the non conditional codings roots are suitable for processors with a fix word length of the coding word, the conditional coding root can be used to model varying bit-widths of the instructions and VLIW-type architectures. In LISA, the description of varying word-lengths is formalized by describing several coding roots. The different coding roots are enumerated by using LISA control structure (switch statement). However, unlike the usage of SWITCH-CASE statements embracing sections in operations which can switch either group or label values, the SWITCH-CASE statement used to express multiple coding roots switches an enumeration type. The enumeration type ENUM is declared in the declare section of the operation and lists the names of alternative coding roots. At the time of decoding, the different cases are passed through until a valid instruction is identified.

*cond-coding-root:*
  SWITCH '(' *coding-root-type* ')' '{'

  CASE *coding-root-type*$_1$ ':' '{'
  *coding-root*$_1$
  '}'

  ...

  { CASE *coding-root-type*$_n$ ':' '{'
  *coding-root*$_n$
  '}' }

  '}'

*non-cond-coding-root:*
  *coding-root*

*coding-root:*
  CODING AT '(' *program-counter* ')' '{'
  *insn-comparison*
  '}'

Each particular coding root is defined by the keywords CODING AT, followed by the fetch address (*program-counter*) of the instruction which is processed next. The *program-counter* is enclosed by parentheses.

*insn-comparison:*
  *comp-single-instruction*
  | *comp-diff-word-length*
  | *comp-vliw-word*

There are three cases which have to be distinguished regarding the comparison of the actual instruction word with the coding description in LISA. In the first case a single instruction word is compared to a list of identifiers. These identifiers may be either groups, instances, or labels and are compared with an instruction register (*insn-register*).

*comp-single-instruction:*
  $insn\text{-}register == CodingElement_1 \ldots CodingElement_n$

The second case covers architectures working with multi-word instructions. Here several instruction registers are compared to a list of *CodingElements*. The usage of multiple instruction words is formalized in LISA by concatenating the comparisons of different instruction register contents to the LISA coding tree by a logical AND (&&) operator. The respective comparisons are enclosed in round braces.

*comp-diff-word-length:*
  '(' $insn\text{-}register_1 == CodingElement_1 \ldots CodingElement_m$ ')'
  ...
  { && '(' $insn\text{-}register_n == CodingElement_1 \ldots CodingElement_k$ ')' }

The third case deals with specialities found in VLIW architectures. To cope with the problem to decode VLIW packets, LISA allows splitting up these packets into smaller pieces, which are logically connected in the coding root by using the OR (||) operator. In this case the assembler treats each sub-instruction independently and considers it within the address calculation. This enables the use of symbolic addresses (labels) in context with sub-instructions of a VLIW packet. The OR operators (||) between the assignments indicate that all assignments belong to a single instruction word.

*comp-vliw-word:*
 '(' $insn\text{-}register_1$ == $Group_1$ ')'
 ...
 { || '(' $insn\text{-}register_n$ == $Group_1$ ')' }

*insn-register:*
 *ResourceElement*

## Coding-Section

The coding section provides information, which is used in two manners: for decoding and thus, for the selection of LISA operations by comparing to the respective bits in the binary instruction word, and for assembling the appropriate binary instruction words. The bit widths of instructions within a LISA model must be constant. That means, that for all permutations (over groups) of the coding section, the sum of the element bit-sizes must not change. The syntax of a coding section is given below.

*CodingSection:*
 CODING '{' *CodingElement* { *CodingElement* } '}'

A coding section is introduced by the keyword CODING and, enclosed in curly braces, followed by a list of coding elements. One coding element can either be a group, an instance, or a terminal coding like a label or a bit-field. In case of an *instance*, the coding of the corresponding operation is referenced. A *group* specifies a subset of LISA operations' coding allowed in the current context. The member of this group is selected by the decoder, which matches the respective binary instruction portion of the operation. The selected member operation can be used to evaluate a LISA control-structure to create a dependency of the other sections on the coding. A single coding element is defined as follows.

*CodingElement:*
 *group-name* [ '[' *start* .. *end* ']' ]
 | *instance-name* [ '[' *start* .. *end* ']' ]
 | *Bitfield*
 | *label-name* = *Bitfield*
 | *label-name* = *Arithmetic Expression*

Besides symbolic coding elements, terminal binary coding elements are allowed. *Binary coding* is specified by a sequence of *0*s, *1*s and *x*s, preceded by the prefix *0b* to indicate binary format. Each *0* and *1* embodies a single bit of the instruction word. An *x* describes a *don't care* bit, thus, the value is ignored during decoding. The syntax of a binary coding element is depicted below.

*Bitfield:*
`0b [ 0 | 1 | x ]`
| `0b0` '[' *bit-size* ']'
| `0b1` '[' *bit-size* ']'
| `0bx` '[' *bit-size* ']'

Terminal coding elements can be assigned to labels, which are placeholders for numeric values. The assignment is performed by the specification of the `label-name` followed by an equal sign and a bit field element. LISA enables the developer to perform several arithmetic operations with the coding fragment, before it is assigned to the label. The implemented operators for use in such expressions are: add (+), subtract (-), and left respectively right shift (<<,>>).

*Arithmetic Expression:*
*Bitfield* | *Expression*

*Expression:*
'(' *Operation* ')'

*Operation:*
*Operand* | *Expression coding-operator Expression*

*Operand:*
*Bitfield* | '(' [ + | - ] *Operand* ')'

*coding-operator:*
+ | -- | << | >>

**Distributed Coding:** To obtain maximum flexibility, the coding referenced by a group or an instance can be distributed over the complete instruction word. This is a case frequently found in low-power architectures to get a highly compact coding word (e.g. ARM7). Here, the coding for a group or instance is not coherent in the coding section – that is, for example a group refers to the first two and the last three bits of a coding section. Consequently the position of a distributed coding element must be specified. The *start* and *end* parameters in square brackets behind the group or instance name within a coding section define the destination position. The length is fixed by these parameters. No overlapping of code fragments in the destination element is allowed, and the sum of the fragment sizes must match the bit-size of the entire destination element.

## Syntax-Section

The syntax section contains the assembly syntax of an operation. The syntax is needed for assembling and disassembling purposes. In contrast to the coding section, one operation may have have multiple syntax sections which can be either exclusive or sequentially concatenated. This can be achieved by making several syntax sections conditional (via the LISA control-structure) and others not. Thus, a single operation can describe a huge amount of syntax permutations. The assembly syntax is again composed by several syntax elements. As in the coding section, a syntax element can be either a group, a reference, an instance, or a terminal. Groups, references, and instances reference the syntax section of the respective operations which must contain a syntax section themselves.

*SyntaxSection:*
    SYNTAX '{'
    [ ~ ] *SyntaxElement*$_1$
    ...
    { [ ~ ] *SyntaxElement*$_n$ }
    }

The '~ ' sign specifies that there is no white-space allowed between the current and the previous syntax element. Besides, string fragments in double quotes are allowed in a syntax section. These string fragments represent the terminal syntax information. Leading white-spaces of a string fragment are ignored unless the '~ ' sign is used. The LISA syntax of a syntax element is defined as follows:

*SyntaxElement:*
    *group-name*
    | *instance-name*
    | '' *string* ''
    | *FormatedLabel*
    | *GenericLabel*

As in the coding-section, it is possible to use labels within the syntax-section. A label requires some extra format specifications for the disassembly generation and the assembly parsing. A sign extension can be performed on the label by the use of the #S format specifier followed by the position of the sign. The output format can be specified as well: hexadecimal (#X), binary (#B), unsigned (#U) and signed decimal (#S) formatting is supported. The output-format specification is needed for the disassembler to generate the appropriate assembly. If the signed output format is chosen, a postfix is mandatory to mark the sign bit of the element. This is done by the *SignPosition* which is given as a *decimal-number*.

*FormatedLabel:*
    *Label Expression* = *OutputFormat*

*Label Expression:*
    *label-name*
    | ( *label-name* = InputFormat )
    | *Arithmetic Expression*

*OutputFormat:*
    #X | #B | #U | #S[ *SignPosition* ]

*InputFormat:*
    #S[ *SignPosition* ]

The output-format of a label specifies the expected number format in the assembly code. Thus, it is not allowed to use a hexadecimal value instead of a binary where a binary output format was specified. A number prefix or postfix, as required for hexadecimal values, is not included in a LISA label. It has to be added explicitly by a leading (prefix) or ending (postfix) string fragment (like "0x" or "h").

A special item related within the syntax-section applies additional properties to labels. By the use of the SYMBOL keyword, a label can represent any number format, symbolic addresses, and arithmetic expressions. This is e.g. required for convenient modelling of branch instructions.

The restriction to a fixed address at compile-time of the LISA model would be an unacceptable limitation. Using symbols, also relocatable code is supported. However, the output-format for the disassembler can still be specified, including a pre- or postfix. The syntax for these generic labels is as follows :

*GenericLabel:*
  `SYMBOL '(' [''`*prefix string*`'' ]` *FormatedLabel* `[''`*postfix string*`'' ] ')'`

## Behavior-Section

The behavior section provides information about the state update functions. The functionality of the current operation is written in conventional C/C++ code. Moreover, the behavior function of other operations can be called via groups, instances, or references. This is performed by using the respective identifier with trailing parenthesis plus a final semicolon, equal to ordinary C-function calls. Besides calling the behavior functions of other operations, regular C/C++-functions which are defined outside the LISA description may be called. These can either reside in pre-compiled libraries or user-defined C/C++-source files, which are added to the LISA description. Furthermore, the expression-sections of other LISA operations can be referenced simply by using the identifier of groups, instances, or references which contain operations declaring expressions. Theses expressions contain resources defined in the resource section of the LISA description. All processor resources defined in the resource-section of the LISA model are global by definition, and thus can be accessed and modified by any function or behavior code. This has the great advantage that the used resources do not have to be provided as parameters in a function call.

*Behavior Section:*
  `BEHAVIOR '{'` *BehaviorCode* `'}'`

*BehaviorCode:*
  *BehaviorCode* | *CStatement* | *BehaviorCall*

*BehaviorCall:*
  *group-name* `'(' ')' ';'` | *operation-name* `'(' ')' ';'`

## Activation-Section

Within the activation section the timing behavior of other operations relative to the current operation is defined. The timing information is specified in two ways. Either timing is determined from the spatial ordering of pipeline stages which operations are assigned to. Besides, the operation timing can be affected by the use of commas or semicolons in the activation-section. While the use of the semicolon adds a delay cycle to the execution, using the comma to separate activated operations reflects concurrency. The syntax of the activation section is given below.

*Activation Section:*
`ACTIVATION` '{' *ActivationList* '}'

*ActivationList:*
*ActivationElement*$_1$ [ ',' ... ',' *ActivationElement*$_n$ ]
| *ActivationList If_Activation*
| *ActivationList Else_Activation*

*ActivationElement:*
*group-name* | *operation-name* | *PipeFunction*

Conditional activation is provided by means of *if-else* statements. Here, an ordinary C-like expression represents the condition which may include processor resources or references to expression sections of other operations.

*If_Activation:*
`if` '(' *cond-expression* ')' *ActivationList*

*Else_Activation:*
`else` *ActivationList*

## Expression-Section

The expression-section is used for indirect resource and immediate value access in a LISA processor model. As with any other section, the expression-section may contain references to other operations via groups, instances, or references. The referenced operations have to contain an expression-section as well. The syntax of the expression-section is given below.

*Expression Section:*
`EXPRESSION` '{' *CastedExpression* '}'

Type casts are provided in the same way as in the programming language C. Therefore it is possible to explicitly cast the whole expression to any native data type of the C language or a user-defined type.

*CastedExpression:*
*Expression* | '(' *expression-type* ')' *Expression*

*Expression:*
*ExpressionElement*
| *LabelOrNumber expr-operator LabelOrNumber*
{ *expr-operator LabelOrNumber* }

Additionally, several arithmetic operators are supported to allow embedding of the label elements into arithmetic expressions. The supported operators are given below.

*ExpressionElement:*
*ResourceElement* | *group-name* | *operation-name* | *LabelOrNumber*

*LabelOrNumber:*
*label-name* | *number*

*expr-operator:*
+ | -- | * | / | << | >>

## LISA Control Structure

The LISA language provides control-structures to create a dependency of the syntax, behavior, activation etc. on the coding and syntax of an operation.

Two types of control-structures are supported by LISA: IF/ELSE and SWITCH/CASE constructs. The evaluated expression can be either a label value or a group/reference state. The state of a group or reference refers to the member operation of groups/references which is selected when decoding/assembling. The syntax is defined as follows:

*IF-ELSE Section:*
`IF` '(' *condition*')' `THEN` '{' *section list* '}' `ELSE` '{' *section list* '}'

*condition:*
*statement* | '(' *condition* ')' | '(' *condition conjunction condition*')'

*statement:*
*label-name cts-operator number*
*group-name cts-operator operation-name*
*reference-name cts-operator operation-name*

*cts-operator:*
== | !=

*conjunction:*
&& | ||

*SWITCH-CASE Section:*
`SWITCH` '(' *label-name* | *group-name* | *reference-name* )
'{' $case_1$ { $case_n$ } [ *default case* ] '}'

*case:*
`CASE` *number* | *operation-name* ':' '{' *section list* '}'

*default case:*
`DEFAULT` ':' '{' *section list* '}'

All sections of a LISA operation can be conditional. Depending on the binary instruction word , the respective syntax, behavior, expression, or activation sequence is selected.

# Appendix C
# Sample ARM7 LISA Model

In the following chapter, a LISA 2.0 model of the ARM7 processor core from Advanced Risc Machines Ltd. (ARM) is presented. It was derived from the publicly available information on the architecture which can be found on the the ARM web-page under *http://www.arm.com* and in the ARM Architecture Reference Manual by David Seal [192]. The presented model corresponds to the one used for measuring the speed of the generated simulation tools in chapter 7.4.1.1.

The LISA 2.0 processor model is divided into several files which are organized by the categories of instructions that are described:

- *main.lisa* contains the description of processor resources and the reserved LISA operations `main` and `reset`.
- *decode.lisa* contains the description of the coding-root which splits the model into different instruction classes. The following files are all related to the description of these instruction classes.
- *data_proc.lisa* describes the instructions modelling the data-transfer from the PSR register to the general register-set.
- *data_swap* models single data swaps in the ARM7 architecture.
- *multiply.lisa* describes the multiplication of two values.
- *single_trans.lisa* describes load and store operations of single values to and from memory.
- *block_trans.lisa* models the load and store operations of several values to and from memory.
- *branch.lisa* contains the description of instructions modelling control flow.
- *soft_int.lisa* models software interrupts in the processor.
- *coproc.lisa* contains instructions steering the co-processor that can be attached to the ARM7. In the LISA model, the description of a co-processor is omitted, however, the model is already prepared to include arbitrary C-code to model the behavior of a co-processor.
- *misc.lisa* comprises miscellaneous LISA 2.0 operations needed to complete the model.

The realized LISA 2.0 model is instruction-based and thus contains no description of the underlying micro-architecture such as pipelines. The reason for that is founded in the fact that

ARM does not make micro-architectural information of their architectures publicly available. However, it would be quite simple and straightforward to enrich the model accordingly.

Due to limited space in this book, the model has been shortened compared to the original version. To obtain the complete model, please contact *info@lisatek.com.*

## Resources and reserved operations main and reset – file main.lisa

```
#include "arm7.h"

/***************************************************************/
/*    Declaration of the ARM7100 resources                     */
/***************************************************************/

RESOURCE
{
  MEMORY_MAP
    {
      // Memory is accessed bytewise, model memory is organized wordwise
      BUS(bus_pmem), RANGE(0x00000000,0x000fffff) -> prog_mem[(31..2)];
      BUS(bus_dmem), RANGE(0x00100000,0x001fffff) -> data_mem[(31..2)];
      BUS(bus_heap), RANGE(0x40000000,0x40800000) -> heap_mem[(31..2)];
      BUS(bus_stck), RANGE(0x7fff8000,0x7fffffff) -> stack_mem[(31..2)];
      BUS(bus_peri), RANGE(0x80000000,0x8000ffff) -> peripheral_data_mem[(31..2)];
    }

  // Buses
  BUS U32 bus_pmem
  {
     ADDRTYPE(U32);
     BLOCKSIZE(32,8);
  };
  ...

  // Memory
  RAM U32 data_mem
  {
     ADDRTYPE(U32);
     BLOCKSIZE(32,8);
     ENDIANESS(LITTLE);
     SIZE( 0x40000 );
     FLAGS(R|W|X);
  };
  ...

  //
  // General Registers all used in USER32 mode
  // R15 is always the program counter
  //
  REGISTER
    U32 R[0..15],
    R_fiq[8..14],              // Special Register for FIQ32 mode
    R_svc[13..14],             // Special Register for SUPERVISOR32 mode
    R_abt[13..14],             // Special Register for ABORT32 mode
    R_irq[13..14],             // Special Register for IRQ32 mode
    R_und[13..14];             // Special Register for UNDEFINED32 mode

  U32_PTR R_mode[0..15];
  U32_PTR SPSR_mode;

  //
  // Status register
  //

  // Bit 31 = N status bit (negative/less than)
  // Bit 30 = Z status bit (zero)
```

```
  // Bit 29 = C status bit (carry/borrow/extend)
  // Bit 28 = V status bit (overflow)
  // Bit 27 - 8 = unused
  // Bit 7  = I flag -> IRQ disabled
  // Bit 6  = F flag -> FIQ disabled
  // Bit 5  = unused
  // Bit 4  = M4 mode bit
  // Bit 3  = M3 mode bit
  // Bit 2  = M2 mode bit
  // Bit 1  = M1 mode bit
  // Bit 0  = M0 mode bit
  CONTROL_REGISTER
    U32 CPSR,              // Current Program Status Register
    SPSR_fiq,              // Saved Program Status Registers (banked), FIQ32 mode
    SPSR_svc,              // Saved Program Status Registers (banked), SUPERVISOR32 mode
    SPSR_abt,              // Saved Program Status Registers (banked), ABORT32 mode
    SPSR_irq,              // Saved Program Status Registers (banked), IRQ32 mode
    SPSR_und,              // Saved Program Status Registers (banked), UNDEFINED32 mode
    SPSR_dummy;            //     pointed to by SPSR_mode in User mode, but never written to.

  U32   super_mode_info;            // storage of comments contained in SWI instruction

  // register register reads register writes
  //       ---------------------------------------------
  // ctrl[0] ID reserved
  // ctrl[1] reserved control
  // ctrl[2] reserved translation table base
  // ctrl[3] reserved domain access control
  // ctrl[4] reserved reserved
  // ctrl[5] fault status flush tlb
  // ctrl[6] fault status purge tlb
  // ctrl[7] reserved flush idc
  // ctrl[8..15] reserved reserved
  U32 ctrl[0..15];

  //
  // External PINs
  //

  // PINs indicating interrupts
  PIN bool   nRESET;    // RESET PIN, highest priority
  PIN bool   nFIQ;      // Fast Interrupt PIN
  PIN bool   nIRQ;      // Normal Interrupt PIN
  PIN bool   ABORT;

  // Endianess -> LOW(0) is little, HIGH(1) is big endian
  U32   BIGEND;

  // PROG32 and DATA32 -> used for backward compability
  U32   PROG32, DATA32;

  PROGRAM_COUNTER U32 PC ALIAS R[15];     // Mapped on R[15]
}


/*************************************************************/
/*    RESET                                                  */
/*************************************************************/
OPERATION reset
{
  BEHAVIOR
    {
      // Set base registers to USER32-mode
      for(int n = 0; n < 16; n++)
       R_mode[n] = &R[n];

      //
      // Zero all resources
```

```
      //
      for(int i = 0; i < 16; i++)
        R[i] = 0;
      for(int j = 8; j < 15; j++)
        R_fiq[j] = 0;
      for(int k = 13; k < 15; k++)
        R_irq[k] = R_svc[k] = R_abt[k] = R_irq[k] = R_und[k] = 0;

      CPSR = SPSR_fiq = SPSR_svc = SPSR_abt = SPSR_irq = SPSR_und = SPSR_dummy = 0;

      ....
}


/************************************************************/
/*    MAIN OPERATION                                        */
/*                                                          */
/*    this is the root process that initiates new           */
/*    instructions                                          */
/************************************************************/
OPERATION main
{
  DECLARE
    {
      INSTANCE Decode_Instruction, Interrupt_Detection;
    }
  BEHAVIOR
    {
      cycle++;
      mode_switch=0; // flag mode_switch prevents that mode is switched twice a cycle.

      // Load instruction into register
      LOAD_WORD(PC,insn_register);

      // Execute instruction
      Decode_Instruction();

      // Interrupt detection

      // Interrupts take affect with the next instruction
      // -> thus they are checked after the current instruction
      // is executed
      if (!mode_switch) Interrupt_Detection();

      // Increment program counter
      PC += 4;
      }
    }
}


/************************************************************/
/*    INTERRUPT DETECTION                                   */
/*                                                          */
/*    This operation checks the occurance of interrupts     */
/************************************************************/
OPERATION Interrupt_Detection
{
  BEHAVIOR
    {
      //
      // Reset has highest priority
      //
      if(!nRESET)
        {
          R_svc[14] = R[15];
          SPSR_svc = CPSR;
          SET_SVC_MODE;
```

```
        BANK_SVC_MODE;
        SET_MASK_FIQ;
        SET_MASK_IRQ;

        PC = 0x0-4;
      }
    //
    // Check for FIQ Interrupt (Fast interrupt request)
    //

    // Interrupt is requested when pin nFIQ is low and not masked out
    else if(!nFIQ && !GET_MASK_FIQ)
      {
  ....

  }
}
```

# Description of the decoder – file decode.lisa

```
#include "arm7.h"

/**************************************************************/
/*                                                            */
/*   Instruction decoder                                      */
/*                                                            */
/**************************************************************/
OPERATION Decode_Instruction
{
  DECLARE
    {
      GROUP instruction_type = {
         Data_Processing_PSR_Transfer || Data_Processing_ALU || Multiply ||
         Data_Processing_CMPU || Single_Data_Swap || Single_Data_Transfer || Undefined ||
         Block_Data_Transfer || Branch || Coproc_Data_Transfer ||
         Coproc_Data_Operation || Coproc_Register_Transfer ||
         Software_Interrupt };
      GROUP cond = {
      EQ || NE || CS || CC || MI || PL || VS || VC || HI || LS ||
      GE || LT || GT || LE || NV || AL };
    }

  CODING AT (PC & 0xfffffffc) { insn_register == cond instruction_type  }
  SYNTAX { instruction_type }
  BEHAVIOR { instruction_counter++; }

  // Operations are only executed if condition evaluates true
  SWITCH(cond) {
      CASE EQ: {
        BEHAVIOR
          {
            if(GET_Z_FLAG)
              instruction_type();
          }
      }
      CASE NE: {
        BEHAVIOR
          {
            if(!GET_Z_FLAG)
               instruction_type();
          }
        }
      ...

    DEFAULT: {
        BEHAVIOR
          { /* Instruction never executed */ }
```

```
        }
    }
}
```

## Description of data processing PSR transfer – file data_proc.lisa

```
#include "arm7.h"

/**************************************************************/
/*    Data_Processing_PSR_Transfer                            */
/*                                                            */
/*    Instruction copies from and to PSR register             */
/*    (Special encoding of instructions TEQ, TST,             */
/*    CMN and CMP without the S bit set)                      */
/**************************************************************/
ALIAS OPERATION Data_Processing_PSR_Transfer
{
  DECLARE {
      REFERENCE cond;
      GROUP MRS_or_MSR = { MRS || MSR }; }
  CODING { MRS_or_MSR }
  SYNTAX { MRS_or_MSR }
  BEHAVIOR { MRS_or_MSR(); }
}


//
// Transfer PSR contents to a register
//
OPERATION MRS
{
  DECLARE {
      REFERENCE    cond;
      GROUP dest = { reg };
      GROUP Ps   = { is_CPSR || is_SPSR || is_CPSR_all || is_SPSR_all };
    }
  CODING { 0b00010 Ps 0b001111 dest 0b0[12] }
  IF((Ps == is_CPSR) || (Ps == is_CPSR_all)) THEN {
      SYNTAX {
      "MRS" ~cond dest ~"," Ps
    }
      BEHAVIOR {
      *dest = CPSR;
    }
    }
  ELSE {
      SYNTAX {
      "MRS" ~cond dest ~"," Ps
    }
      BEHAVIOR {
      *dest = *SPSR_mode;
    }
    }
}


//
// Transfer to PSR
//
OPERATION MSR
{
  DECLARE {
      REFERENCE          cond;
      GROUP reg_or_imm = { MSR_reg || MSR_imm }; }
  CODING { reg_or_imm }
  SYNTAX { reg_or_imm }
  BEHAVIOR { reg_or_imm(); }
```

```
}


// Transfer register contents to PSR
OPERATION MSR_reg
{
  DECLARE {
      REFERENCE   cond;
      GROUP Pd =  { is_CPSR || is_SPSR || is_CPSR_all || is_SPSR_all };
      GROUP src = { reg }; }
  CODING { 0b00010 Pd 0b1000011111 0b0[8] src }

  IF((Pd == is_CPSR) || (Pd == is_CPSR_all)) THEN
    {
      SYNTAX { "MSR" ~cond Pd ~"," src }
      BEHAVIOR {
          if (GET_USER_MODE)
          {
            CPSR = (CPSR & 0x0fffffff) | (*src & 0xf0000000);
          } else
          {
            CPSR = *src;
          }
      UPDATE_REGISTER_POINTER;
    }
    }
  ELSE
    {
      SYNTAX { "MSR" ~cond Pd ~"," src }
      BEHAVIOR {
      if (!GET_USER_MODE) *SPSR_mode = *src;
    }
    }
}


// Transfer register contents or immediate value to PSR flag bits only
OPERATION MSR_imm
{
  DECLARE {
      REFERENCE           cond;
      GROUP type_of_imm   = { operand2_is_register || operand2_is_immediate };
      GROUP Pd            = { is_CPSR || is_SPSR };
      INSTANCE            Source_operand;
    }
  CODING {
      0b00 type_of_imm 0b10 Pd 0b1010001111 Source_operand
    }
  SYNTAX {
      "MSR" ~cond Source_operand
    }
  BEHAVIOR { Source_operand(); }
}


// Source operand is register
OPERATION Source_operand
{
  DECLARE {
      REFERENCE      type_of_imm, Pd;
      GROUP src    = { reg };
      LABEL          rot, imm;
    }

  IF(type_of_imm == operand2_is_register) THEN
    {
      CODING { 0b0[8] src }
      IF(Pd == is_CPSR) THEN
```

```
    {
      SYNTAX { "CPSR_flg" ~"," src }
      BEHAVIOR {
          CPSR &= 0x0fffffff;
          CPSR |= (*src & 0xf0000000);
        }
    }
      ELSE
    {
      SYNTAX { "SPSR_flg" ~"," src }
      BEHAVIOR {
          *SPSR_mode &= 0x0fffffff;
          *SPSR_mode |= (*src & 0xf0000000);
        }
    }
    }
  ELSE
    {
      CODING { rot=0bx[4] imm=0bx[8] }
      IF(Pd == is_CPSR) THEN
    {
      SYNTAX {
          "CPSR_flg" ~"," SYMBOL("#" imm=#U) ~"," rot
        }
      BEHAVIOR {
          CPSR &= 0x0fffffff;
          CPSR |= (rot_imm(imm, rot, 0, R, CPSR) & 0xf0000000);
        }
    }
      ELSE
    {
      SYNTAX {
          "SPSR_flg" ~"," SYMBOL("#" imm=#U) ~"," rot
        }
      BEHAVIOR {
          *SPSR_mode &= 0x0fffffff;
          *SPSR_mode |= (rot_imm(imm, rot, 0, R, CPSR) & 0xf0000000);
        }
    }
    }
}


OPERATION is_CPSR
{
  CODING { 0b0 }
  SYNTAX { "CPSR" }
}

ALIAS OPERATION is_CPSR_all
{
  CODING { 0b0 }
  SYNTAX { "CPSR_all" }
}

...

/************************************************************/
/*    Data_Processing_ALU                                   */
/*                                                          */
/*    Instruction produces a result by performing a         */
/*    specified arithmetic or logical operation on one      */
/*    or two operands                                       */
/************************************************************/
OPERATION Data_Processing_ALU
{
  DECLARE {
      REFERENCE cond;
```

```
      GROUP     OpCode      = { AND || EOR || SUB || RSB || ADD || ADC || SBC ||
                        RSC || ORR || MOV || BIC || MVN };
      GROUP     src1, dest  = { reg || R15 };
      GROUP     setcond     = { change_cond_code || leave_cond_code };
      GROUP         operand2        = { op2_reg || op2_imm };
    }
  CODING {
      00b operand2=[12..12] OpCode setcond src1 dest operand2=[0..11]
    }

  SYNTAX { OpCode ~operand2 }
  BEHAVIOR {
      // Determine second operand for ALU instruction
      operand2();

      // Execute instruction -> read second operand from barrel-shifter bus
      OpCode();
    }
}

OPERATION Data_Processing_CMPU
{
  DECLARE {
      REFERENCE cond;
      GROUP     OpCode      = { TST || TEQ || CMP || CMN };
      GROUP     src1, dest  = { reg || R15 };
      GROUP     setcond     = { change_cond_code };
      GROUP         operand2        = { op2_reg || op2_imm };
    }
  CODING {
      00b operand2=[12..12] OpCode setcond src1 dest operand2=[0..11]
    }
  SYNTAX { OpCode ~operand2 }
  BEHAVIOR {
      // Determine second operand for ALU instruction
      operand2();

      // Execute instruction -> read second operand from barrel-shifter bus
      OpCode();
    }
}


// Second operand is a shifted register
OPERATION op2_reg
{
  DECLARE {
      REFERENCE     src1, dest, OpCode, setcond, cond;
      GROUP     src2 = { reg };
      GROUP         shift_param_type = { shift_param_reg || shift_param_imm };
  }
  CODING { 0b0 shift_param_type src2 }
  SYNTAX { shift_param_type }
  BEHAVIOR { shift_param_type(); }
}

....
```

## Description of single data swap operation – file data_swap.lisa

```
#include "arm7.h"


/****************************************************************/
/*    Single Data Swap                                          */
/****************************************************************/
OPERATION Single_Data_Swap
```

```
{
  DECLARE {
      REFERENCE cond;
      GROUP     src, dest, base_reg = { reg };
      GROUP swap_type= { swap_byte || swap_word };
    }
  CODING {
      0b00010 swap_type 0b00 base_reg dest 0b0000 0b1001 src
    }
  SYNTAX {
      "SWP" ~swap_type dest ~"," src ~",[" ~base_reg ~"]"
    }

  IF (swap_type == swap_word) THEN
    {
      BEHAVIOR {
          LOAD_WORD(*dest,*base_reg);
          STORE_WORD(*src,*base_reg);
      }
    }
  ELSE
    {
      BEHAVIOR {
          LOAD_BYTE(*dest,*base_reg);
          STORE_BYTE(*src,*base_reg);
      }
    }
}


OPERATION swap_byte
{
  DECLARE { REFERENCE cond; }
  CODING { 0b1 }
  SYNTAX { cond ~"B" }
}


OPERATION swap_word
{
  DECLARE { REFERENCE cond; }
  CODING { 0b0 }
  SYNTAX { cond }
}
```

## Description of the multiplication operation – file multiply.lisa

```
#include "arm7.h"

/************************************************************/
/*    Multiply                                              */
/************************************************************/
OPERATION Multiply
{
  DECLARE {
      REFERENCE cond;
      GROUP  setaccumulate = { setmac || setmul };
      GROUP setcond = { change_cond_code || leave_cond_code };
      INSTANCE multiply_parameter;
    }
  CODING {
      0b0[6] setaccumulate setcond multiply_parameter
    }
  SYNTAX {
      setaccumulate ~setcond multiply_parameter
    }
  BEHAVIOR { multiply_parameter(); }
```

```
}


// multiply and accumulate
OPERATION setmac
{
  CODING { 0b1 }
  SYNTAX { "MLA" }
}


// multiply only
OPERATION setmul
{
  CODING { 0b0 }
  SYNTAX { "MUL" }
}


//
// Perform multiplication or MAC
//
OPERATION multiply_parameter
{
  DECLARE {
      REFERENCE setaccumulate, setcond;
      GROUP dest, srcadd, src1, src2 = { reg };
    }
  CODING { dest srcadd src2 0b1001 src1 }
  IF (setaccumulate == setmac) THEN
    {
      SYNTAX { dest ~"," src1 ~"," src2 ~"," srcadd }
      BEHAVIOR {
          // CXBit muss hier noch rein ....
          *dest = *src1 * *src2 + *srcadd;

          // Change condition register CPSR
          UPDATE_CPSR_LU(*dest,setcond);
      }
    }
  ELSE
    {
      SYNTAX { dest ~"," src1 ~"," src2 }
      BEHAVIOR {
          // CXBit muss hier noch rein ....
          *dest = *src1 * *src2;

          // Change condition register CPSR
          UPDATE_CPSR_LU(*dest,setcond);
      }
    }
}
```

## Description of a single data transfer – file single_trans.lisa

```
#include "arm7.h"

/****************************************************************/
/*    Load/Store Single Data Transfer                           */
/****************************************************************/
OPERATION Single_Data_Transfer
{
  DECLARE {
      REFERENCE     cond;
      GROUP         type_of_offset = { offset_is_imm || offset_is_reg };
      GROUP         write_back     = { is_write_back || is_not_write_back };
      GROUP         up_down        = { is_up || is_down };
```

```
      GROUP         byte_word = { is_byte || is_word };
      GROUP         pre_post  = { is_pre  || is_post };
      GROUP         up_down   = { is_up   || is_down };
      GROUP         write_back= { is_write_back || is_not_write_back };
      GROUP         instruction = { load || store };
      GROUP         src_dest_reg, base_reg = { reg || R15 };
      INSTANCE      offset;
    }
  CODING {
      0b01 type_of_offset pre_post up_down byte_word write_back
        instruction base_reg src_dest_reg offset
    }

  // Depending on the value of the write-back and pre_post bits,
  // different syntax is used (with or without letter T attached
  // to mnemonic)
  IF ((write_back == is_write_back) && (pre_post == is_post)) THEN
    {
      SYNTAX {
          instruction ~byte_word ~"T" src_dest_reg ~"," offset
        }
    }
  ELSE
    {
      SYNTAX {
          instruction ~byte_word ~" " src_dest_reg ~"," offset
        }
    }

  // First: calculate offset_value and write result on offset_value_bus
  // Second: Execute load/store in dependence of the bits write_back,
  // pre_post and up_down
  IF (pre_post == is_pre) THEN
    {
      BEHAVIOR {
          offset();
          write_back();
          instruction();
        }
    }
  ELSE
    {
      BEHAVIOR {
          offset();
          up_down();
          instruction();
          write_back(); // write back bit is redundant.
        }
    }
}


OPERATION offset
{
  DECLARE {
      REFERENCE       type_of_offset, pre_post, write_back, up_down, base_reg,
                      src_dest_reg, instruction, byte_word;
      LABEL           immediate;
      GROUP           src2 = { reg };
      GROUP           shift_param_type =  { transfer_shift_param_imm };
    }

  //
  // Offset is immediate value
  //
  IF (type_of_offset == offset_is_imm) THEN
    {
      CODING { immediate=0bx[12] }
```

```
      BEHAVIOR {
          // Put immediate value on offset bus
          offset_bus = immediate;
        }
      IF (write_back == is_write_back) THEN
        {
          // SYNTAX: pre_indexed version with immediate
          IF (pre_post == is_pre) THEN
            {
              SYNTAX {
                  "[" ~base_reg ~"," up_down SYMBOL("#" immediate=#X) ~"]" "!"
                }
            }
          // SYNTAX: post_indexed version with immediate
          ELSE
            {
              SYNTAX {
                  "[" ~base_reg ~"]," up_down SYMBOL("#" immediate=#X)  "!"
                }
            }
        }
      ELSE
        {
          IF(base_reg == R15) THEN
            {

      ....
}


//
//
// Offset is a register -> shift value of offset register is specified as immediate
//
// Shift value of offset register contained in a register as used in data_processing
// is not available with this instruction
//
OPERATION transfer_shift_param_imm
{
  DECLARE {
      REFERENCE           instruction, up_down, src_dest_reg, base_reg, src2,
                          pre_post, write_back, byte_word;
      LABEL               shift_amount;
      GROUP               shift_type = { logical_left || logical_right || arithmetic_right ||
                                      arithmetic_left || rotate_right };
  }
  CODING { shift_amount=0bx[5] shift_type 0b0 }
  SYNTAX { src2 }

  IF (shift_amount != 0) THEN
  {
    SYNTAX { ~"," shift_type SYMBOL("#" shift_amount=#X) }
  }

  // Evaluate shifted register value and write result on offset_bus
  SWITCH (shift_type)
    {
      CASE logical_left: {
        BEHAVIOR {
            offset_bus = imm_lsl(*src2, shift_amount, 0, R, CPSR);
          }
      }
      CASE logical_right: {
        BEHAVIOR {
            offset_bus = imm_lsr(*src2, shift_amount,0, R, CPSR);
          }
      }
      .....
```

```
}


//
// Write back enabled
//
OPERATION is_write_back
{
  DECLARE { REFERENCE base_reg, up_down; }
  CODING { 0b1 }
  SYNTAX { "!" }
  IF(up_down == is_up) THEN
    {
      BEHAVIOR {
          *base_reg += offset_bus;
          address_bus = *base_reg;
      }
    }
  ELSE
    {
      BEHAVIOR {
          *base_reg -= offset_bus;
           address_bus = *base_reg;
      }
    }
}

....


OPERATION load
{
  DECLARE { REFERENCE byte_word, src_dest_reg; }
  CODING { 0b1 }
  SYNTAX { "LDR" }
  // load byte
  IF(byte_word == is_byte) THEN
    {
      BEHAVIOR {
          LOAD_BYTE(address_bus,*src_dest_reg);
        }
    }
  // load word
  ELSE
    {
      BEHAVIOR {
          LOAD_WORD(address_bus,*src_dest_reg);
      }
    }
  // special treatment if PC is reloaded !
  IF (src_dest_reg == R15) THEN
    {
      BEHAVIOR { (*src_dest_reg)-=4; }
    }
}


OPERATION store
{
  DECLARE { REFERENCE byte_word, src_dest_reg; }
  CODING { 0b0 }
  SYNTAX { "STR" }
  // write byte
  IF (byte_word == is_byte) THEN
    {
      BEHAVIOR {
          STORE_BYTE(address_bus,*src_dest_reg);
        }
```

```
    }
  // write word
  ELSE
    {
      BEHAVIOR {
         STORE_WORD(address_bus,*src_dest_reg);
        }
    }
}
```

# Description of a block data transfer – file block_trans.lisa

```
#include "arm7.h"

/*************************************************************/
/*    Load/Store Block Data Transfer                         */
/*************************************************************/
bOPERATION Block_Data_Transfer
{
  DECLARE {
      REFERENCE cond;
      GROUP  instruction = { block_load || block_store };
      GROUP pre_post = { block_pre_indexing_yes || block_pre_indexing_no };
      GROUP up_down = { block_up || block_down };
      GROUP psr = { load_psr || dont_load_psr };
      GROUP  write_back = { write_back_ok || write_back_not_ok };
      GROUP        base_reg = { reg || R13_stackptr };
      INSTANCE          reglist;
    }
  CODING {
      0b100 pre_post up_down psr write_back instruction base_reg reglist
    }
  IF(base_reg == R13_stackptr) THEN
    {
      SYNTAX {
  instruction ~cond ~pre_post ~up_down base_reg ~write_back ~","
    ~"{" reglist "}" ~psr
}
      BEHAVIOR { reglist(); }
    }
  ELSE
    {
      SYNTAX {
  instruction ~cond ~up_down ~pre_post base_reg ~write_back ~","
    ~"{" reglist "}" ~psr
}
      BEHAVIOR { reglist(); }
    }
}


OPERATION reglist
{
  DECLARE {
      REFERENCE  write_back, pre_post, instruction, up_down, base_reg, psr;
      LABEL bit_reg0, bit_reg1, bit_reg2, bit_reg3, bit_reg4, bit_reg5,
       bit_reg6, bit_reg7, bit_reg8, bit_reg9, bit_reg10, bit_reg11,
       bit_reg12, bit_reg13, bit_reg14, bit_reg15;
    }
  CODING {
      bit_reg15=0bx bit_reg14=0bx bit_reg13=0bx bit_reg12=0bx bit_reg11=0bx
       bit_reg10=0bx bit_reg9=0bx bit_reg8=0bx bit_reg7=0bx bit_reg6=0bx
       bit_reg5=0bx bit_reg4=0bx bit_reg3=0bx bit_reg2=0bx bit_reg1=0bx
       bit_reg0=0bx
    }

  //
```

```
// Put register to load/store on virtual stack
//
  IF(bit_reg0 == 1) THEN
    {
      SYNTAX { "R0 " }
    }
  IF(bit_reg1 == 1) THEN
    {
      IF(bit_reg0 == 1) THEN
       {
         SYNTAX { "," "R1 " }
       }
      ELSE
       {
         SYNTAX { "R1 " }
       }
    }
  IF(bit_reg2 == 1) THEN
    {
      IF((bit_reg0 == 1) || (bit_reg1 == 1)) THEN
       {
         SYNTAX { "," "R2 " }
       }
      ELSE
       {
         SYNTAX { "R2 " }
       }
    }
  ....

  IF(bit_reg15 == 1) THEN
    {
      IF((bit_reg0 == 1) || (bit_reg1 == 1) || (bit_reg2 == 1) || (bit_reg3 == 1) ||
        (bit_reg4 == 1) || (bit_reg5 == 1) || (bit_reg6 == 1) || (bit_reg7 == 1) ||
        (bit_reg8 == 1) || (bit_reg9 == 1) || (bit_reg10 == 1) || (bit_reg11 == 1) ||
        (bit_reg12 == 1) || (bit_reg13 == 1) || (bit_reg14 == 1)) THEN
       {
         SYNTAX { "," "R15 " }
       }
      ELSE
       {
         SYNTAX { "R15 " }
       }
    }

//
// BEHAVIOR valid for all
//

// read from current register bank
IF ((psr == dont_load_psr) || ((bit_reg15 == 1) && (instruction == block_store))) THEN
  {
    BEHAVIOR {
       U32_PTR save_register[16];
       int current_pos = 0;

       U32 base_register = *base_reg;

       // Put registers on Stack
       if(bit_reg0) save_register[current_pos++] = R_mode[0];
       if(bit_reg1) save_register[current_pos++] = R_mode[1];
       if(bit_reg2) save_register[current_pos++] = R_mode[2];
       if(bit_reg3) save_register[current_pos++] = R_mode[3];
       if(bit_reg4) save_register[current_pos++] = R_mode[4];
       if(bit_reg5) save_register[current_pos++] = R_mode[5];
       if(bit_reg6) save_register[current_pos++] = R_mode[6];
       if(bit_reg7) save_register[current_pos++] = R_mode[7];
       if(bit_reg8) save_register[current_pos++] = R_mode[8];
```

```
        if(bit_reg9) save_register[current_pos++] = R_mode[9];
        if(bit_reg10) save_register[current_pos++] = R_mode[10];
        if(bit_reg11) save_register[current_pos++] = R_mode[11];
        if(bit_reg12) save_register[current_pos++] = R_mode[12];
        if(bit_reg13) save_register[current_pos++] = R_mode[13];
        if(bit_reg14) save_register[current_pos++] = R_mode[14];
        if(bit_reg15) save_register[current_pos++] = R_mode[15];
      }
  }
// User bank transfer
ELSE
  {
    BEHAVIOR
      {
        U32_PTR save_register[16];
        int current_pos = 0;

        U32 base_register = *base_reg;

        // Put registers on Stack
        if(bit_reg0) save_register[current_pos++] = &R[0];
        if(bit_reg1) save_register[current_pos++] = &R[1];
        if(bit_reg2) save_register[current_pos++] = &R[2];
        if(bit_reg3) save_register[current_pos++] = &R[3];
        if(bit_reg4) save_register[current_pos++] = &R[4];
        if(bit_reg5) save_register[current_pos++] = &R[5];
        if(bit_reg6) save_register[current_pos++] = &R[6];
        if(bit_reg7) save_register[current_pos++] = &R[7];
        if(bit_reg8) save_register[current_pos++] = &R[8];
        if(bit_reg9) save_register[current_pos++] = &R[9];
        if(bit_reg10) save_register[current_pos++] = &R[10];
        if(bit_reg11) save_register[current_pos++] = &R[11];
        if(bit_reg12) save_register[current_pos++] = &R[12];
        if(bit_reg13) save_register[current_pos++] = &R[13];
        if(bit_reg14) save_register[current_pos++] = &R[14];
        if(bit_reg15) save_register[current_pos++] = &R[15];
      }
  }

// No writeback -- only pre-indexing
IF(write_back == write_back_not_ok) THEN
  {
    // pre
    IF(pre_post == block_pre_indexing_yes) THEN
     {
       // increment
       IF(up_down == block_up) THEN
         {
           BEHAVIOR {
              for(int i=0; i < current_pos; i++)
                {
                  reg = save_register[i];
                  address_bus = base_register + (4*i) + 4;
                  instruction();
                }
            }
         }
  ....

  }
// Writeback -- both pre- and post-indexing
ELSE
  {
    // pre
    IF(pre_post == block_pre_indexing_yes) THEN
    {
  ....
  }
```

```
  ....
}


OPERATION load_psr
{
  CODING { 0b1 }
  SYNTAX { ~"^" }
}


OPERATION dont_load_psr
{
  CODING { 0b0 }
  SYNTAX { ~"" }
}


OPERATION write_back_ok
{
  CODING { 0b1 }
  SYNTAX { ~"!" }
}

....

}
```

## Description of control flow – file branch.lisa

```
/***************************************************************/
/*    Branches                                                 */
/***************************************************************/

OPERATION Branch
{
  DECLARE {
      REFERENCE cond;
      GROUP branch_type = { branch_with_link || branch_without_link };
      LABEL offset;
    }
  CODING { 0b101 branch_type offset=0bx[24] }
  SYNTAX  {
      branch_type ~cond SYMBOL("#0x" ((CURRENT_ADDRESS + ((offset=#S24) << 2)) + 8)=#X)
    }

  // If link bit is set, write old PC into R[14]
  IF(branch_type == branch_with_link) THEN
    {
      BEHAVIOR { *R_mode[14] = PC + 4; }
    }
  BEHAVIOR {
      // The branch offset must take account of the prefetch operation,
      // which causes the PC to be 2 words (8 bytes) ahead of the current
      // instruction
      PC += SIGN_EXT32((offset << 2),26) + 4;
    }
}


OPERATION branch_with_link
{
  CODING { 0b1 }
  SYNTAX { "BL" }
}
```

```
OPERATION branch_without_link
{
  CODING { 0b0 }
  SYNTAX { "BN" }
}
```

# Description of software interrupts – file soft_int.lisa

```
#include "arm7.h"

/************************************************************/
/*    Software Interrupt (SWI)
/************************************************************/
OPERATION Software_Interrupt
{
  DECLARE {
      LABEL           comment;
      REFERENCE       cond;
    }
  CODING { 0b1111 comment=0bx[24] }
  SYNTAX { "SWI" ~cond SYMBOL("#" comment=#U) }
  BEHAVIOR {
      if (comment!=0x123456)
       { // if user SWI are used with SVC mode already being active, the user has to save registers
         R_svc[14] = R[15] + 4;
         SPSR_svc = CPSR;
         BANK_SVC_MODE;
         SET_SVC_MODE;
         SET_MASK_IRQ;
         PC = 0x08-4; // will be incremented to 0x08 later.
       }
      else
       {

    .....

    }
}
```

# Description of coprocessor instructions – file coproc.lisa

```
#include "arm7.h"

/************************************************************/
/*    Coprocessor Instruction                               */
/*    Data Operations                                       */
/************************************************************/
OPERATION Coproc_Data_Operation
{
  DECLARE {
      REFERENCE   cond;
      LABEL       CP_Opc, CRn, CRd, CP_num,CP, CRm;
    }
  CODING {
      0b1110 CP_Opc=0bx[4] CRn=0bx[4] CRd=0bx[4] CP_num=0bx[4]
      CP=0bx[3] 0b0 CRm=0bx[4]
    }
  SYNTAX {
      "CDP" ~cond CP_num=#U "not implemented" CP_Opc=#U CRn=#U CRd=#U
      CP=#U CRm=#U
    }
  BEHAVIOR { /* not implemented */  }
}
```

```
/************************************************************/
/*    Coprocessor Instruction                               */
/*    Data Transfer                                         */
/************************************************************/
OPERATION Coproc_Data_Transfer
{
  DECLARE {
      REFERENCE   cond;
      LABEL       P, U, N, W, Rn, CRd, CP_num, Offset;
      GROUP L    = { co_load || co_store };
    }
  CODING {
      0b110 P=0bx U=0bx N=0bx W=0bx L Rn=0bx[4]
      CRd=0bx[4] CP_num=0bx[4] Offset=0bx[8]
    }
  IF(L == load) THEN
    {
      SYNTAX {
        "LDC" ~cond CP_num=#U "not implemented" P=#U U=#U N=#U W=#U L=#U Rn=#U
          CRd=#U Offset=#U L
      }
    }
  ELSE
    {
      SYNTAX {
        "STC" ~cond CP_num=#U "not implemented" P=#U U=#U N=#U W=#U L=#U Rn=#U
          CRd=#U Offset=#U L
      }
    }
  BEHAVIOR { }
}


/************************************************************/
/*    Coprocessor Instruction                               */
/*    Register Transfer                                     */
/************************************************************/
OPERATION Coproc_Register_Transfer
{
  DECLARE {
      REFERENCE   cond;
      LABEL       CPOpc, CRn, Rd, CP_num, CP, CRm;
      GROUP L    = { co_load || co_store };
    }
  CODING {
      0b1110 CPOpc=0bx[3] L CRn=0bx[4] Rd=0bx[4] CP_num=0bx[4]
      CP=0bx[3] 0b1 CRm=0bx[4]
    }
  IF(L == load) THEN
    {
      SYNTAX {
        "MCR" ~cond CP_num=#U "not implemented" CPOpc=#U CRn=#U
          Rd=#U CP=#U CRm=#U L
      }
    }
  ELSE
    {
      SYNTAX {
        "MRC" ~cond CP_num=#U "not implemented" CPOpc=#U CRn=#U
          Rd=#U CP=#U CRm=#U L
      }
    }
  BEHAVIOR { }
}


OPERATION co_load
{
```

```
  CODING { 0b1 }
  SYNTAX { "LDR" }
  BEHAVIOR { }
}


OPERATION co_store
{
  CODING { 0b0 }
  SYNTAX { "STR" }
  BEHAVIOR { }
}
```

## Description of miscellaneous operations – file misc.lisa

```
#include "arm7.h"

/***************************************************************/
/*    Register set                                             */
/***************************************************************/
OPERATION reg
{
  DECLARE { LABEL index; }
  CODING { index=0bx[4] }
  SYNTAX { "R" ~index=#U }
  EXPRESSION { (U32_PTR) R_mode[index] }
}


ALIAS OPERATION R13_stackptr
{
  CODING { 0b1101 }
  SYNTAX { "R13" }
  EXPRESSION { (U32_PTR) R_mode[13] }
}


ALIAS OPERATION R15
{
  CODING { 0b1111 }
  SYNTAX { "R15" }
  EXPRESSION { (U32_PTR) R_mode[15] }
}


/**********************************************************************/
/*        Undefined Instruction                                       */
/**********************************************************************/
OPERATION Undefined
{
  DECLARE { LABEL undef1, undef2; }
  CODING { 0b011 undef1=0bx[20] 0b1 undef2=0bx[4] }
  SYNTAX { "UNDEFINED" undef1=#U undef2=#U }
  BEHAVIOR {
      // Save address of undefined instruction
      R_und[14] = R[15] + 4;
      SET_UND_MODE;
      SET_MASK_IRQ;
      R[14] = 0x04;
    }
}


/**********************************************************************/
/*        Condition fields (bits 31..28) in each ARM7 instruction     */
/**********************************************************************/
```

```
// equal - Z set
OPERATION EQ
{
  CODING { 0b0000 }
  SYNTAX  { ~"EQ" }
}


// not equal - Z clear
OPERATION NE
{
  CODING { 0b0001 }
  SYNTAX { ~"NE" }
}

....
```

# Appendix D
# The ICORE Architecture

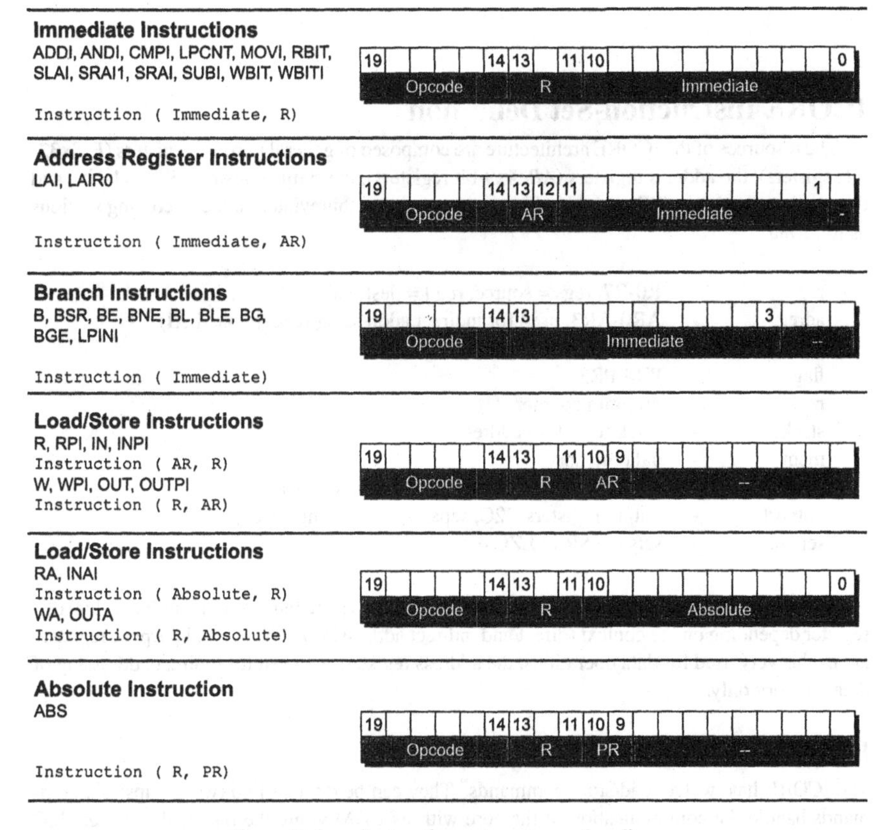


*Figure D.1a.* Instruction-set syntax and coding.

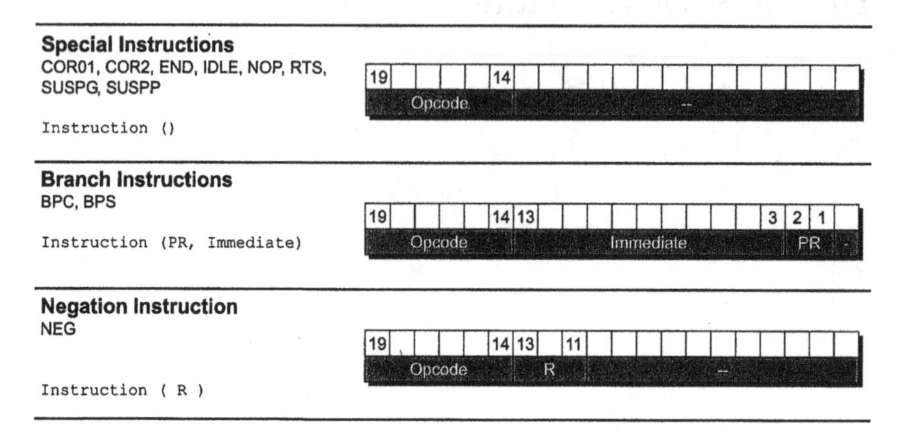


*Figure D.1b.* Instruction-set syntax and coding.

# ICORE Instruction-Set Definition

The resources of the ICORE architecture are composed of general purpose registers (*R*, 8x32-bit registers), the address registers (*AR*, 4x9-bit registers), the status registers (*SR*, 3x1-bit), and the predicate registers (PR, 4x1-bit). These resources are abbreviated in the upcoming sections as followed:

| | | |
|---|---|---|
| reg | - | R0-R7, regs = source, regd = destination |
| adreg | - | AR0-AR3, used for indirect addressing (exception: LAI) |
| con | - | 11 bit constant (immediate value) |
| flag | - | PR0-PR3 |
| pc | - | program counter |
| stack | - | stack for return address |
| mem | - | data memory |
| inport | - | input registers (I2C, separate addressing space) |
| outport | - | output registers (I2C, separate addressing space) |
| set_status() | - | sets the SR in LZC order |

*Note:* The mnemonic *adreg* can denote both an address register and the content (i.e. value) of a register depending on the context (direct and indirect addressing). The general purpose registers are exclusively used for data operations, the address registers are used for indirect addressing of data memory only.

## Load/Store Commands

The ICORE has twelve load/store commands. They can be divided into two groups: six commands handle the communication of the core with its RAM while the rest deals with the I2C registers. Table D.1 depicts the different commands.

*Table D.1.* Load/store commands of the ICORE.

| *Command* | *Description* | *Expression* |
|---|---|---|
| R(adreg, reg) | Read memory at address adreg and store value in register reg | $reg = mem(adreg)$ |
| RA(abs, reg) | Read memory at address abs and store value in register reg | $reg = mem(abs)$ |
| RPI(adreg, reg) | Read memory at address adreg, store value in register reg, and post increment adreg | $reg = mem(adreg++)$ |
| W(reg, adreg) | Write register reg to memory at address adreg | $mem(adreg) = reg$ |
| WA(reg, abs) | Write register reg to memory at address abs | $mem(abs) = reg$ |
| WPI(reg, adreg) | Write register reg to memory at address adreg and post increment adreg | $mem(adreg++) = reg$ |
| IN(adreg, reg) | Read input port addressed by adreg and store value in reg | $reg = inport(adreg)$ |
| INA(abs, reg) | Read input port addressed by abs and store value in register reg | $reg = inport(abs)$ |
| INPI(adreg, reg) | Read input port addressed by adreg, store value in register reg, and post increment adreg | $reg = inport(adreg++)$ |
| OUT(reg, adreg) | Write register reg to output port addressed by adreg | $outport(adreg) = reg$ |
| OUTA(reg, abs) | Write register reg to output port addressed by abs | $outport(abs) = reg$ |
| OUTPI(reg, adreg) | Write register reg to output port addressed by adreg and post increment adreg | $outport(adreg++) = reg$ |

## Arithmetic Commands – Arithmetic Commands with Two Registers

*Table D.2a.* Arithmetic commands with two registers.

| *Command* | *Description* | *Expression* |
|---|---|---|
| ADD(regs, regd) | Add regs and regd, store result in regd | $regd = regs + regd$ |
| ADDSUB0(regs, regd) | Add or subtract regs and regd and store result in regd | $(R1 >= 0)?\ (regd+ = regs) : (regd- = regs)$ |
| ADDSUB1(regs, regd) | Subtract or add regs and regd and store result in regd | $(R1 >= 0)?\ (regd- = regs) : (regd+ = regs)$ |
| CMP(regs, regd) | Compare two registers | $set\_status(regd - regs)$ |

*Table D.2b.* Arithmetic commands with two registers.

| *Command* | *Description* | *Expression* |
|---|---|---|
| MULS(regs, regd) | Signed multiplication of the lower 16 bit of regs and regd. Result is stored in regd | $regd = (short)(regd\&65535) * (short)(regs\&65535)$ |
| MULU(regs, regd) | Unsigned multiplication of the lower 16 bit of regs and regd. Result is stored in regd | $regd = (unsigned)(regd\&65535) * (unsigned)(regs\&65535)$ |
| SLA(regs, regd) | Shift left arithmetic (# of shifts is specified by regs' lower 5 bit) | $regd = regd << regs\_low5$ |
| SRA(regs, regd) | Shift right arithmetic (# of shifts is specified by regs' lower 5 bit) | $regd = regd >> regs\_low5$ |
| SRA1(regs,regd) | Special instruction for CORDIC algorithm (shift by nr+1 bits, with nr=zolp_curr_count with final + 0.5 rounding) | $(zolp\_curr\_count <= -1)? (regd = regs) : (regd = (((regs >> \backslash zolp\_curr\_count) +1) >> 1))$ |
| SUB(regs, regd) | Subtract two registers and store in register regd | $regd = regd - regs$ |

## Arithmetic Commands – Arithmetic Commands with Immediate Value

*Table D.3.* Arithmetic commands with an immediate value.

| *Command* | *Description* | *Expression* |
|---|---|---|
| ADDI(con, reg) | Add an immediate value to register reg | $reg = reg + con$ |
| CMPI(con, reg) | Compare a register with an immediate value | $set\_status(reg - con)$ |
| SAT(con, reg) | Limit value of reg to range from $-2^{con}$ to $2^{con} - 1$ | $reg = reg$, if in range, else sat. to max. or min. range value |
| SLAI(con, reg) | Shift left arithmetic with immediate value (# of shifts is specified by con) | $reg = reg << con$ |
| SRAI(con, reg) | Shift right arithmetic with immediate value (# of shifts is specified by con) | $reg = reg >> con$ |
| SRAI1(con, reg) | Special instruction for CORDIC algorithm (shift by con bits with 1 final + 0.5 rounding) | $(con <= -1)?(reg = reg) : (reg = (((reg >> con) + 1) >> 1))$ |
| SUBI(con, reg) | Subtract an immediate value from a register and store in reg | $reg = reg - con$ |

## Arithmetic Commands – Other Arithmetic Commands

*Table D.4.* Other arithmetic commands.

| *Command* | *Description* | *Expression* |
|---|---|---|
| ABS(reg, flag) | Calculate absolute value of reg and store sign in PR(flag) | $(reg < 0)?flag = 1 : flag = 0$ $reg = abs(reg)$ |
| COR01 | Special CORDIC instruction | $(R1 >= 0)?(R3+ = R6) : (R3- = R6);$ $R5 = R7;$ $(R1 >= 0)?(R2+ = R4) : (R2- = R4)$ $(zolp\_curr\_count == -1)?$ $(R7 = R2) : (R7 = ((R2 >> \backslash$ $zolp\_curr\_count) + 1) >> 1)$ $R6 = mem(AR0 + +)$ |
| COR2 | Special CORDIC instruction | $(R1 < 0)?(R1+ = R5) : (R1- = R5);$ $(zolp\_curr\_count == -1)?$ $(R4 = R1) : (R4 = ((R1 >> \backslash$ $zolp\_curr\_count) + 1) >> 1)$ |
| NEG(reg) | Negate register value | $reg = -reg$ |

## Branch Instructions

*Table D.5a.* Branch Instructions.

| *Command* | *Description* | *Expression* |
|---|---|---|
| B(con) | Unconditional relative branch | $pc+ = con$ |
| BCC(con) | Relative branch if "carry" condition in SR is clear | $(status.c == 0)?$ $pc+ = con$ |
| BCS(con) | Relative branch if "carry" condition in SR is set | $(status.c == 1)?$ $pc+ = con$ |
| BE(con) | Relative branch if "zero" condition in status register | $(status.z == 1)?$ $pc+ = con$ |
| BGE(con) | Relative branch if "greater-or-equal" condition in SR | $(status.l == 0 \| \| status.z == 0)?$ $pc+ = con$ |
| BGT(con) | Relative branch if "greater-than" condition in SR | $(status.l == 0\&\&status.z == 0)?$ $pc+ = con$ |
| BLE(con) | Relative branch if "less-or-equal" condition in SR | $(status.l == 1 \| \| status.z == 1)?$ $pc+ = con$ |
| BLT(con) | Relative branch if "less-than" condition in SR | $(status.l == 1\&\&status.z == 0)?$ $pc+ = con$ |
| BNE(con) | Relative branch if "not-equal" condition in SR | $(status.z == 0)?$ $pc+ = con$ |
| BPC(flag, con) | Relative branch if predicate bit in PR(flag) is clear | $(PR(flag) == 0)?$ $pc+ = con$ |

*Table D.5b.* Branch Instructions.

| *Command* | *Description* | *Expression* |
|---|---|---|
| BPS(flag, con) | Relative branch if predicate bit in PR(flag) is set | $(PR(flag) == 1)?$ $pc+ = con$ |
| BSR(con) | Unconditional relative branch to subroutine | $stack = pc;$ $pc+ = con$ |

## Program Control Instructions

*Table D.6.* Program control instructions.

| *Command* | *Description* | *Expression* |
|---|---|---|
| END | Enforces PPU to go into "idle" mode | – |
| LPCNT(con, reg) | Initialize loop count register | $zolp\_end\_count = reg;$ $zolp\_curr\_count = con$ |
| LPINI(con) | Initialize loop start and end address | $zolp\_start\_adr = pc + 1;$ $zolp\_end\_adr = pc + con$ |
| RTS | Return from subroutine | $pc = stack$ |
| SUSPG | Wait until guard_trig = '1' | $(guard\_trig == 1)?\ ppu\_state = busy$ |
| SUSPP | Wait until ppubus_en = '1' | $(ppubus\_en == 1)?\ ppu\_state = busy$ |

## Miscellaneous Instructions

*Table D.7a.* Miscellaneous instructions.

| *Command* | *Description* | *Expression* |
|---|---|---|
| LAI(con, adreg) | Load address register immediate | $adreg = con$ |
| LAIR0(con, adreg) | Load address register immediate with displacement in R0 | $adreg = con + R0$ |
| MOV(regs, regd) | Move register from source to destination | $regd = regs$ |
| MOVI(con, reg) | Move imm value to register | $reg = con$ |
| RBIT(con, reg) | Write bits from left_position= con_right+con_length-1 to right_position=con_right of register reg in con_length LSBs of register R0 | $R0 = ((reg >> con\_right)\&$ $((1 << con\_length) - 1));$ $con = con\_length * 8$ $+con\_right$ |

*Table D.7b.* Miscellaneous instructions.

| *Command* | *Description* | *Expression* |
|---|---|---|
| WBIT(con, reg) | Replace bits from left_pos.= con_right+con_length-1 to right_pos.=con_right in register R0 with con_length LSBs in register reg | $R0 = (((R0 >> (con\_left + 1)) << (con\_left) + 1)) + ((reg\&\backslash ((1 << con\_left - con\_right + 1))\backslash -1)) << con_right) + (R0\& ((1 << con\_right) - 1)));$ $con = con\_length * 8 + con\_right$ |
| WBITI(con, reg) | Replace bits from left_pos.= con_right+con_length-1 to right_pos.=con_right in register reg with con_length LSBs of con_value | $reg = (((reg >> (con\_left + 1))\backslash << (con\_left + 1)) + ((31\&\backslash con\_value\&((1 << con\_left - \backslash con\_right + 1)) - 1)) << con\_right) + (reg\&((1 << con\_right) - 1)));$ $con = 64 * con\_value + 8* con\_length + con\_right$ |

## Power Consumption Charts

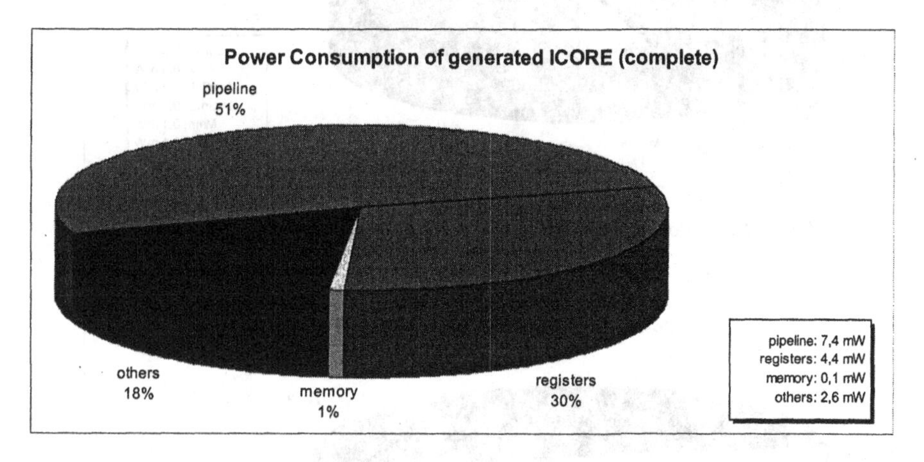


*Figure D.2.* Power consumption of the generated ICORE (complete architecture).

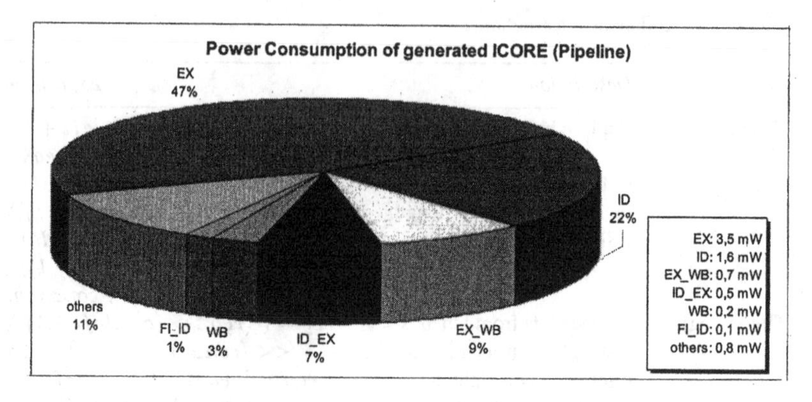


*Figure D.3.* Power consumption of the generated ICORE model (pipeline only).

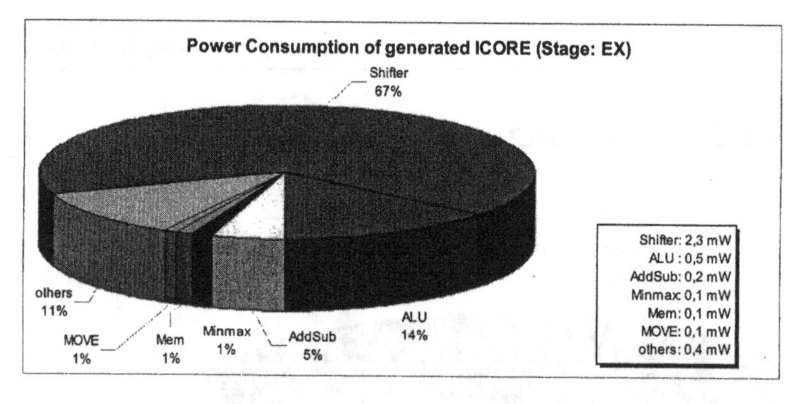


*Figure D.4.* Power consumption of the generated ICORE model (stage: EX).

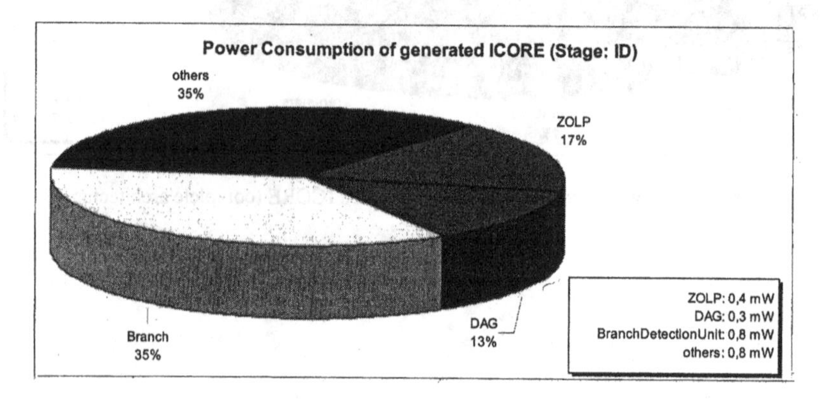


*Figure D.5.* Power consumption of the generated ICORE model (stage: ID).

```
PEQ:
  MOVI(24,R7);       // load end value of loop counter
  LAI (12,AR1);      // load pointer to register with ppu bus word 0
  INPI(AR1,R2);
  INPI(AR1,R1);
  LAI (0,AR0);       // set pointer to start of cordic angle tab
  MOVI(0,R3);        // angle := 0
  CMPI(0,R2);
  NOP;
  BNE(L1);
  CMPI(0,R1);
  NOP;
  BE(CORDIC_END);
L1:
  ABS(R2,PR0);
  ABS(R1,PR1);       // store sign of y in predicate sy
  NOP;
  NOP;
  MOV(R1,R4);
  MOV(R2,R5);
  RPI(AR0,R6);       // read angle increment da from table
  LPCNT(0,R7);       // initialize dedicated loop count register to start value
  NOP;               // of zero and end value R7 (for following SRA1)
  ADDSUB0(R6,R3);
  ADDSUB0(R4,R2);
  ADDSUB1(R5,R1);    // R1 update must be last!!
  RPI(AR0,R6);
  NOP;
  SRA1(R1,R4);       // ATTENTION: SRA1 implicitly takes dedicated loop counter
  SRA1(R2,R5);       // as shift count */
  LPCNT(1,R7);       // initialize dedicated loop count register to start value
  NOP;               // of zero and end value R7
  MOV(R5,R7);
  LPINI(COR_LOOPSTART,COR_LOOPEND); // initialize loop start and loop end register
COR_LOOPSTART:
  NOP;
  COR01;             // ADDSUB0(R6,R3); MOV(R7,R5); ADDSUBSH0(R4,R2,R7); RPI(AR0,R6)
  NOP;
  NOP;
  NOP;
  COR2;              // ADDSUBSH1(R5,R1,R4);
  NOP;
  NOP;
COR_LOOPEND:
  BPC(PR0, L3);
  MOVI(1,R4);
  MOVI(23,R7);
  NOP;
  NOP;
  SLA(R7,R4);        // R4=0.5
  NOP;
  NOP;
  SUB(R3,R4);        // R4=0.5-R3
  NOP;
  NOP;
  MOV(R4,R3);
L3:
  BPC(PR1, CORDIC_END);        // branch if PR1==0 to end of cordic
  NEG(a);
CORDIC_END:
  LAI(0,AR1);                  // initialize output pointer
  NOP;
  OUTPI(R2,AR1);
  OUTPI(R1,AR1);
  OUTPI(R3,AR1);
  NOP;
  END_INSTRUCTION;
```

*Example D.1:* Test-program used to gather power consumption values.

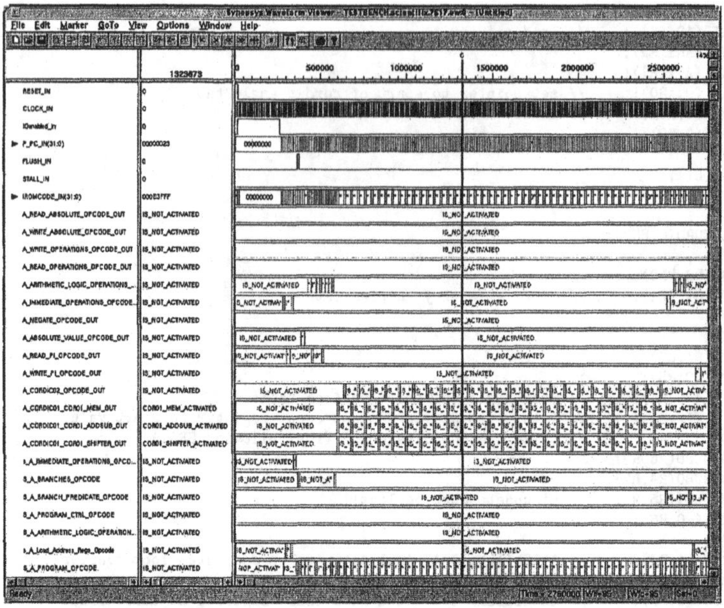

*Figure D.6.* Synopsys waveform viewer for the generated ICORE VHDL model.

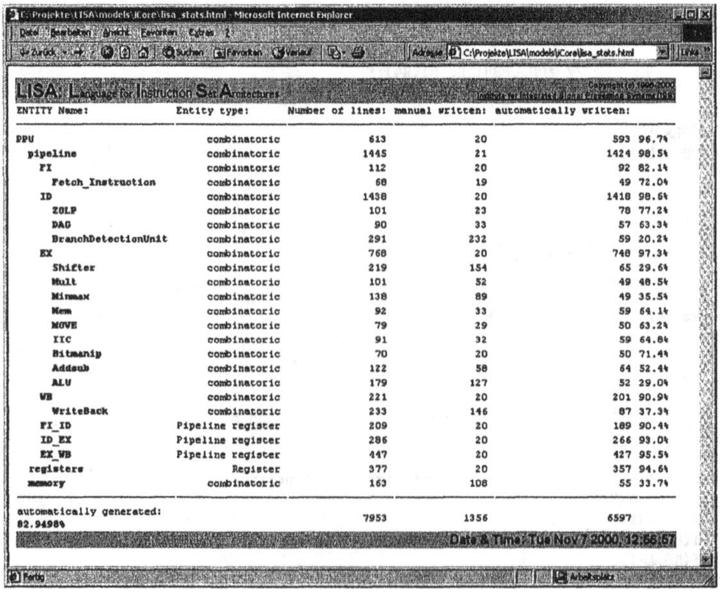

LISA: Language for Instruction Set Architectures

| ENTITY Name: | Entity type: | Number of lines: | manual written: | automatically written: | |
|---|---|---|---|---|---|
| PPU | combinatoric | 613 | 20 | 593 | 96.7% |
| pipeline | combinatoric | 1445 | 21 | 1424 | 98.5% |
| FI | combinatoric | 112 | 20 | 92 | 82.1% |
| Fetch_Instruction | combinatoric | 68 | 19 | 49 | 72.0% |
| ID | combinatoric | 1438 | 20 | 1418 | 98.6% |
| ZOLP | combinatoric | 101 | 23 | 78 | 77.2% |
| DAG | combinatoric | 90 | 33 | 57 | 63.3% |
| BranchDetectionUnit | combinatoric | 291 | 232 | 59 | 20.2% |
| EX | combinatoric | 768 | 20 | 748 | 97.3% |
| Shifter | combinatoric | 219 | 154 | 65 | 29.6% |
| Mult | combinatoric | 101 | 52 | 49 | 48.5% |
| Minmax | combinatoric | 138 | 89 | 49 | 35.5% |
| Mem | combinatoric | 92 | 33 | 59 | 64.1% |
| MOVE | combinatoric | 79 | 29 | 50 | 63.2% |
| IIC | combinatoric | 91 | 32 | 59 | 64.8% |
| Bitmanip | combinatoric | 70 | 20 | 50 | 71.4% |
| Addsub | combinatoric | 122 | 58 | 64 | 52.4% |
| ALU | combinatoric | 179 | 127 | 52 | 29.0% |
| WB | combinatoric | 221 | 20 | 201 | 90.9% |
| WriteBack | combinatoric | 233 | 146 | 87 | 37.3% |
| FI_ID | Pipeline register | 209 | 20 | 189 | 90.4% |
| ID_EX | Pipeline register | 286 | 20 | 266 | 93.0% |
| EX_WB | Pipeline register | 447 | 20 | 427 | 95.5% |
| registers | Register | 377 | 20 | 357 | 94.6% |
| memory | combinatoric | 163 | 108 | 55 | 33.7% |
| automatically generated: 82.9498% | | 7953 | 1356 | 6597 | |

Date & Time: Tue Nov 7 2000, 12:56:57

*Figure D.7.* Statistics page generated from the LISA processor compiler.

# List of Figures

# List of Examples

# List of Tables

# References


[1] M. Birnbaum and H. Sachs, "How VSIA Answers the SOC Dilemma," *IEEE Computer*, vol. 32, pp. 42–50, June 1999.

[2] Forward Concepts – Electronic Market Research, *http://www.fwdconcepts.com*, 2001.

[3] C. Herring, "Microprocessors, Microcontrollers, and Systems in the New Millenium," *IEEE Micro*, vol. 20, pp. 45–51, Nov. 2000.

[4] M. Flynn and P. Hung and K. Rudd, "Deep-Submicron Microprocessor Design Issues," *IEEE Micro*, vol. 19, pp. 11–22, July/Aug. 1999.

[5] G. Frantz, "Digital Signal Processor Trends," *IEEE Micro*, vol. 20, pp. 52–59, Nov. 2000.

[6] G. Frantz, "SOC – A System Perspective," *IEEE International Electron Devices Meeting (IEDM)*, Dec. 1999.

[7] P. Marwedel, "Code Generation for Core Processors," in *Proc. of the Design Automation Conference (DAC)*, Jun. 1997.

[8] E. A. Lee, "Programmable DSP Architectures: Part I," *IEEE ASSP Magazine*, vol. 5, pp. 4–19, Oct. 1988.

[9] E. A. Lee, "Programmable DSP Architectures: Part II," *IEEE ASSP Magazine*, vol. 6, pp. 4–14, Jan. 1989.

[10] G. Moore, "Cramming more components onto integrated circuits," *Electronics magazine*, Apr. 1965.

[11] M. Santarini, "ASIPs: Get Ready for Reconfigurable Silicon," *EETimes magazine*, Nov. 2000.

[12] M.J. Bassand C.M. Christensen, "The Future of the Microprocessor Business," *IEEE Micro*, pp. 34–39, Apr. 2002.

[13] C. Shannon, "A mathematical theory of communication," *Bell System Technical Journal*, vol. 27, pp. 379–423 and 623–656, Jun. 1948.

[14] T. Gloekler and S. Bitterlich and H. Meyr, "Increasing the Power Efficiency of Application Specific Instruction Set Processors Using Datapath Optimization," in *Proc. of the IEEE Workshop on Signal Processing Systems (SIPS)*, Oct. 2001.

[15] Texas Instruments, *TMS320C62x/C67x CPU and Instruction Set Reference Guide*, Mar. 1998.

[16] Advanced Risc Machines Ltd., *ARM7100 Data Sheet*, Dec. 1994.

[17] XILINX, *http://www.xilinx.com/products*, 2001.

[18] ALTERA, *http://www.altera.com/products*, 2001.

[19] D.M. Brooks and P. Bose and S.E. Schuster and H. Jacobson and P.N. Kudva and A. Buyuktosunoglu and J.-D. Wellman and V. Zyuban and M. Gupta and P.W. Cook, "Power-Aware Microarchitecture: Design and Modeling Challenges for Next-Generation Microprocessors," *IEEE Micro*, vol. 20, pp. 26–44, Nov. 2001.

[20] SPEC – Standard Performance Evaluation Corporation, *http://www.specbench.com*, 2001.

[21] M. Chiodo and P. Giusto and H. Hsieh and A. Jurecska and L. Lavagno and A. Sangiovanni-Vincentelli, "Hardware-Software Codesign of Embedded Systems ," *IEEE Micro*, pp. 26–36, Aug. 1994.

[22] J. Hou and W. Wolf, "Partitioning methods for hardware-software co-design," in *Proc. of the Int. Workshop on Hardware/Software Codesign*, 1996.

[23] D. Herrmann and J. Henkel and R. Ernst, "An Approach to the Adaptation of Estimated Cost Parameters in the COSYMA System," in *Proc. of the Int. Workshop on Hardware/Software Codesign*, 1994.

[24] P. Paulin and M. Cornero and C. Liem, "Trends in Embedded System Technology: An Industrial Perspective," in *Hardware/Software Co-Design* (M. G. M. Sami, ed.), Kluwer Academic Publishers, 1996.

[25] A. Hoffmann and T. Kogel and A. Nohl and G. Braun and O. Schliebusch and A. Wieferink and H. Meyr, "A Novel Methodology for the Design of Application Specific Instruction Set Processors (ASIP) Using a Machine Description Language," *IEEE Transactions on Computer-Aided Design*, vol. 20, pp. 1338–1354, Nov. 2001.

[26] Synopsys, *CoCentric System Studio*
*http://www.synopsys.com/products/cocentric_studio/cocentric_studio.html*, 2001.

[27] M.K. Jain and M. Balakrishnan and A. Kumar, "ASIP Design Methodologies: Survey and Issues," in *Int. Conf. on VLSI Design*, Jan. 2001.

[28] Institute of Electrical and Electronics Engineers, Inc. (IEEE), "IEEE Standard VHDL Language Reference Manual 2000."

[29] Institute of Electrical and Electronics Engineers, Inc. (IEEE), "IEEE Standard for Verilog Hardware Description Language 2001 ."

[30] Synopsys, *Design Compiler*
*http://www.synopsys.com/products/logic/logic.html*, 2001.

[31] K. Olukotun and M. Heinrich and D. Ofelt, "Digital System Simulation: Methodologies and Examples," in *Proc. of the Design Automation Conference (DAC)*, Jun. 1998.

[32] J. Rowson, "Hardware/Software co-simulation," in *Proc. of the Design Automation Conference (DAC)*, 1994.

[33] R. Leupers and P. Marwedel, "Instruction Set Extraction From Programmable Structures," in *Proc. of the European Design Automation Conference (EuroDAC)*, Mar. 1994.

[34] H. Tomiyama and A. Halambi and P. Grun and N. Dutt and A. Nicolau, "Architecture Description Languages for System-on-Chip Design," in *Proc. of the Asia Pacific Conference on Chip Design Language (APCHDL)*, Oct. 1999.

[35] A. Halambi and P. Grun and H. Tomiyama and N. Dutt and A. Nicolau, "Automatic Software Toolkit Generation for Embedded System-on-Chip," in *Proc. of the International Conference on Visual Computing*, Feb. 1999.

[36] M. Freericks, "The nML Machine Description Formalism," Technical Report, Technical University of Berlin, Department of Computer Science, 1993.

[37] J. van Praet and G. Goossens and D. Lanner and H. De Man, "Instruction Set Definition and Instruction Selection for ASIPs," in *Proc. of the Int. Symposium on System Synthesis (ISSS)*, Oct. 1994.

[38] G. Hadjiyiannis and S. Hanono and S. Devadas, "ISDL: An Instruction Set Description Language for Retargetability," in *Proc. of the Design Automation Conference (DAC)*, Jun. 1997.

[39] A. Inoue and H. Tomiyama and E.F. Nurprasetyo and H. Yasuura, "A Programming Language for Processor Based Embedded Systems," in *Proc. of the Asia Pacific Conference on Chip Design Language (APCHDL)*, 1999.

[40] N. Ramsey and J.W. Davidson, "Machine Descriptions to Build Tools for Embedded Systems," in *Workshop on Languages, Compilers, and Tools for Embedded Systems*, 1998.

[41] S. Bashford and U. Bieker and B. Harking and R. Leupers and P. Marwedel and A. Neumann and D. Voggenauer, "The MIMOLA Language, Version 4.1," Reference Manual, Department of Computer Science 12, Embedded System Design and Didactics of Computer Science, 1994.

[42] T. Morimoto and K. Yamazaki and H. Nakamura and T. Boku and K. Nakazawa, "Superscalar processor design with hardware description language AIDL," in *Proc. of the Asia Pacific Conference on Chip Design Language (APCHDL)*, Oct. 1994.

[43] H. Akaboshi, *A Study on Design Support for Computer Architecture Design*. PhD thesis, Department of Information Systems, Kyushu University, Jan. 1996.

[44] P. Paulin and C. Liem and T.C. May and S. Sutarwala, "FlexWare: A Flexible Firmware Development Environment for Embedded Systems," in *Code Generation for Embedded Processors* (P. Marwedel and G. Goossens, eds.), Kluwer Academic Publishers, 1995.

[45] J.C. Gyllenhaal and W.W. Hwu and B.R. Rau, "Optimization of Machine Descriptions for Efficient Use," *International Journal of Parallel Programming*, Aug. 1998.

[46] J. Sato and M. Imai and T. Hakata and A. Alomary and N. Hikichi, "An Integrated Design Environment for Application-Specific Integrated Processors," in *Proc. of the Int. Conf. on Computer Design (ICCD)*, Mar. 1991.

[47] A. Kitajima and M. Itoh and J. Sato and A. Shiomi and Y. Takeuchi and M. Imai, "Effectiveness of the ASIP Design System PEAS-III in Design of Pipelined Processors," in *Proc. of the Asia South Pacific Design Automation Conference (ASPDAC)*, Jan. 2001.

[48] C. Siska, "A Processor Description Language Supporting Retargetable Multi-Pipeline DSP Program Development Tools," in *Proc. of the Int. Symposium on System Synthesis (ISSS)*, Dec. 1998.

[49] A. Halambi and P. Grun and V. Ganesh and A. Khare and N. Dutt and A. Nicolau, "EXPRESSION: A Language for Architecture Exploration through Compiler/Simulator Retargetability," in *Proc. of the Conference on Design, Automation & Test in Europe (DATE)*, Mar. 1999.

[50] A. Fauth, "Beyond tool-Specific Machine Descriptions," in *Code Generation for Embedded Processors* (P. Marwedel and G. Goossens, eds.), Kluwer Academic Publishers, 1995.

[51] M. Freericks and A. Fauth and A. Knoll, "Implementation of Complex DSP Systems Using High-Level Design Tools," in *Signal Processing VI: Theories and Applications*, 1994.

[52] IMEC, *http://www.imec.be*, 2001.

[53] D. Lanner and J. Van Praet and A. Kifli and K. Schoofs and W. Geurts and F. Thoen and G. Goossens, "Chess: Retargetable Code Generation for Embedded DSP Processors," in *Code Generation for Embedded Processors* (P. Marwedel and G. Goosens, eds.), Kluwer Academic Publishers, 1995.

[54] W. Geurts et "*al.*, "Design of DSP Systems with Chess/Checkers," in *In Proc. of 2nd Int. Workshop on Code Generation for Embedded Processors*, Mar. 1996.

[55] Target Compiler Technologies, *CHESS/CHECKERS*
*http://www.retarget.com*, 2001.

[56] M. Hartoog J.A. Rowson and P.D. Reddy and S. Desai and D.D. Dunlop and E.A. Harcourt and N. Khullar, "Generation of Software Tools from Processor Descriptions for Hardware/Software Codesign," in *Proc. of the Design Automation Conference (DAC)*, Jun. 1997.

[57] V. Rajesh and R. Moona, "Processor Modeling for Hardware Software Codesign," in *Int. Conf. on VLSI Design*, Jan. 1999.

[58] R. Ravindran and R. Moona, "Retargetable Cache Simulation Using High Level Processor Models," in *Proc. of the Computer Security Applications Conference (ACSAC)*, Mar. 2001.

[59] G. Hadjiyiannis, S. Hanono, and S. Devadas, *ISDL Language Reference Manual*, Jan. 1997.

[60] G. Hadjiyiannis and P. Russo and S. Devadas, "A Methodology for Accurate Performance Evaluation in Architecture Exploration," in *Proc. of the Design Automation Conference (DAC)*, Jun. 1999.

[61] A. Inoue, H. Tomiyama, H. Okuma, H. Kanbara, and H. Yasuura, "Language and compiler for optimizing datapath widths of embedded systems," *IEICE Transactions on Fundamentals*, vol. 12, pp. 2595–2604, Dec. 1998.

[62] A. Appel, J. Davidson, and N. Ramsey, "The zephyr compiler infrastructure," internal report, University of Virginia, 1998. http://www.RCS.virginia.edu/zephyr.

[63] N. Ramsey and M.F. Fernandez, "Specifying Representations of Machine Instructions," *IEEE Transactions on Programming Languages and Systems*, vol. 19, Mar. 1997.

[64] M.W. Bailey and J.W. Davidson, "A Formal Model and Specification Language for Procedure Calling Conventions," in *In Proc. of the 22th POPL*, 1998.

[65] R. Leupers and P. Marwedel, "Retargetable Code Compilation Based on Structural Processor Descriptions," *ACM Transactions on Design Automation for Electronic Systems*, vol. 3, pp. 1–36, Jan. 1998.

[66] R. Leupers, *Retargetable Code Generation for Digital Signal Processors*. Kluwer Academic Publishers, 1997.

[67] UDL/I Comittee, *UDL/I Language Reference Manual Version 2.1.0a*, 1994.

[68] B. Moszkowski and Z. Manna, "Reasoning in interval temporal logic," in *Logics of Programs: Proceedings of the 1983 Workshop*, pp. 371–381, Springer-Verlag, 1984.

[69] T. Morimoto and K. Saito and H. Nakamura and T. Boku and K. Nakazawa, "Advanced processor design using hardware description language AIDL," in *Proc. of the Asia South Pacific Design Automation Conference (ASPDAC)*, Mar. 1997.

[70] P. Paulin and J. Frehel and M. Harrand and E. Berrebi and C. Liem and F. Nacabal and J.C. Herluison, "High-Level Synthesis and Codesign Methods: An Application to a Videophone Codec," in *Proc. of the European Design Automation Conference (Euro-DAC)*, Oct. 1995.

[71] P. Paulin and C. Liem and C. May and S. Sutarwala, "CodeSyn: A Retargetable Code Synthesis System," in *Proc. of the Int. Symposium on System Synthesis (ISSS)*, May 1994.

[72] S. Sutarwala and P. Paulin and Y. Kumar, "Insulin: An Instruction Set Simulation Environment," in *Proc. of the CHDL*, Apr. 1993.

[73] P. Paulin, "Towards Application-Specific Architecture Platforms: Embedded Systems Design Automation Technologies," in *Proc. of the EuroMicro*, Apr. 2000.

[74] P. Paulin, "Design Automation Challenges for Application-Specific Architecture Platforms." Keynote speech at SCOPES 2001 - Workshop on Software and Compilers for Embedded Systems (SCOPES), Apr. 2001.

[75] P. Paulin and F. Karim and P. Bromley, "Network Processors: A Perspective on Market Requirements, Processor Architectures and Embedded SW Tools," in *Proc. of the Conference on Design, Automation & Test in Europe (DATE)*, Mar. 2001.

[76] ACE – Associated Compiler Experts, *The COSY compilation system http://www.ace.nl/products/cosy.html*, 2001.

[77] J. Sato and A. Y. Alomary and Y. Honma and T. Nakata and A. Shiomi and N. Hikichi and M. Imai, "PEAS-I: A Hardware/software Codesign System for ASIP Development," *IEICE Transactions on Fundamentals of Electronics, Communications and Computer Sciences*, vol. E77-A, pp. 483–491, Mar. 1994.

[78] R.M. Stallman, *Using and Porting the GNU Compiler Collection.* Free Software Foundation, gcc-2.95 ed., 1999.

[79] S. Kobayashi et "*al.*, "Compiler Generation in PEAS-III: an ASIP Development System," in *Proc. of the Workshop on Software and Compilers for Embedded Systems (SCOPES)*, Mar. 2001.

[80] I.J. Huang and B. Holmer and A.M. Despain, "ASIA: Automatic Synthesis of Instruction-set Architectures," in *Proc. of the SASIMI Workshop*, Oct. 1993.

[81] M. Itoh and S. Higaki and J. Sato and A. Shiomi and Y. Takeuchi A. Kitajima and M. Imai, "PEAS-III: An ASIP Design Environment," in *Proc. of the Int. Conf. on Computer Design (ICCD)*, Sep. 2000.

[82] M. Itoh and Y. Takeuchi and M. Imai and A. Shiomi, "Synthesizable HDL Generation for Pipelined Processors from a Micro-Operation Description," *IEICE Transactions on Fundamentals of Electronics, Communications and Computer Sciences*, vol. E83-A, Mar. 2000.

[83] V. Živojnović and S. Pees and H. Meyr, "LISA – machine description language and generic machine model for HW/SW co-design," in *Proc. of the IEEE Workshop on VLSI Signal Processing*, Oct. 1996.

[84] P. Grun and A. Halambi and A. Khare and V. Ganesh and N. Dutt and A. Nicolau, "EXPRESSION: An ADL for System Level Design Exploration," Tech. Rep. 98-29, Department of Information and Computer Science, University of California, Irvine, Sep. 1998.

[85] A. Khare, "SIMPRESS: A Simulator Generation Environment for System-on-Chip Exploration," tech. rep., Department of Information and Computer Science, University of California, Irvine, Sep. 1999.

[86] A. Khare and N. Savoiu and A. Halambi and P. Grun and N. Dutt and A. Nicolau, "V-SAT: A Visual Specification and Analysis Tool for System-On-Chip Exploration," in *In Proc. of IEEE EUROMICRO*, 1999.

[87] J.-H. Yang and B.-W. Kim and S.-J. Nam and Y.-S. Kwon and D.-H. Lee and J.-Y. Lee and C.-S. Hwang and Y.-H. Lee and S.-H. Hwang and I.-C. Park and C.-M. Kyung, "MetaCore: An Application-Specific Programmable DSP Development System," *IEEE Transactions on Very Large Scale Integration (VLSI) Systems*, vol. 8, pp. 173–183, Apr. 2000.

[88] J.-H. Yang et al., "Metacore: An Application Specific DSP Development System," in *Proc. of the Design Automation Conference (DAC)*, Jun. 1998.

[89] P.M. Kogge, *The Architecture of Pipelined Computers*. Hemisphere Publishing Corp., 1981.

[90] I.J. Huang and A.M. Despain, "Synthesis of Instruction Sets for Pipelined Microprocessors," in *Proc. of the Design Automation Conference (DAC)*, Jun. 1994.

[91] I.J. Huang and A.M. Despain, "Generating Instruction Sets and Microarchitectures from Applications," in *Proc. of the Int. Conf. on Computer Aided Design (ICCAD)*, Nov. 1994.

[92] B.K. Holmer, "A Tool for Processor Instruction Set Design," in *Proc. of the European Design Automation Conference (EuroDAC)*, Apr. 1994.

[93] I.J. Huang and A.M. Despain, "An Extended Classification of Inter-instruction Dependencies and Its Application in Automatic Synthesis of Pipelined Processors," in *In Proc. of the 26th Symposium on Microarchitecture*, Jun. 1993.

[94] I.J. Huang and A.M. Despain, "Hardware/Software Resolution of Pipeline Hazards in Pipeline Synthesis of Instruction Set Processors," in *Proc. of the Int. Conf. on Computer Aided Design (ICCAD)*, Nov. 1993.

[95] I.J. Huang and A.M. Despain, "Synthesis of Application Specific Instruction Sets," *IEEE Transactions on Computer-Aided Design*, vol. 14, Jun. 1995.

[96] V. Zivojnovic, *Der quantitative Ansatz zum gleichzeitigen Entwurf der DSP Architektur und des Compilers*. Dissertation an der RWTH-Aachen: Shaker Verlag, Aachen, 1998. ISBN 3-8265-3919-2.

[97] S. Pees and A. Hoffmann and V. Zivojnovic and H. Meyr, "LISA - Machine Description Language for Cycle-Accurate Models of Programmable DSP Architectures," in *Proc. of the Design Automation Conference (DAC)*, June 1999.

[98] J. Teich and R. Weper, "A Joined Architecture/Compiler Design Environment for ASIPs," in *Proc. of the Conference on Compilers, Architectures and Synthesis for Embedded Systems (CASES)*, Nov. 2000.

[99] J. Teich and R. Weper and D. Fischer and S. Trinkert, "BUILDABONG: A Rapid Prototyping Environment for ASIPs," in *Proc. of the DSP Germany (DSPD)*, Oct. 2000.

[100] R. Woudsma, "EPICS, a Flexible Approach to Embedded DSP Cores," in *Proc. of the Int. Conf. on Signal Processing Applications and Technology (ICSPAT)*, Oct. 1994.

[101] H. Choi and J.H. Yi and J.Y. Kee and I.C. Park and C.M. Kyung, "Exploiting Intellectual Properties in ASIP Designs for Embedded DSP Software," in *Proc. of the Design Automation Conference (DAC)*, Jun. 1999.

[102] H. Choi and J.S. Kim and C.W. Yoon and I.C. Park and S.H. Hwang and C.M. Kyung, "Synthesis of Application Specific Instructions for Embedded "DSP Software," *IEEE Transactions on Computers*, vol. 48, no. 6, pp. 603–614, 1999.

[103] R.J. Cloutier and D.E. Thomas, "Synthesis of Pipelined Instruction Set Processors," in *Proc. of the Design Automation Conference (DAC)*, Jun. 1993.

[104] Y.G. Kim and T.G. Kim, "A Design and Tools Reuse Methodology for Rapid Prototyping of Application Specific Instruction Set Processors," in *In Proc. of the Workshop on Rapid System Prototyping (RSP)*, Apr. 1999.

[105] B.R. Rau and M.S. Schlansker, "Embedded Computer Architecture and Automation," *IEEE Computer*, vol. 34, pp. 75–83, Apr. 2001.

[106] Y. Bajot and H. Mehrez, "Customizable DSP Architecture for ASIP Core Design," in *Proc. of the IEEE Int. Symposium on Circuits and Systems (ISCAS)*, May 2001.

[107] A.S. Terechko and E.J.D. Pol and J.T.J. van Eijndhoven, "PRMDL: a Machine Description Language for Clustered VLIW Architectures," in *Proc. of the Conference on Design, Automation & Test in Europe (DATE)*, Mar. 2001.

[108] B. Shackleford and M. Yasuda and E. Okushi and H. Koizumi and H. Tomiyama and H. Yasuura, "Satsuki: An Integrated Processor Synthesis and Compiler Generation System," in *IEICE Transactions on Information and Systems*, pp. 1373–1381, 1996.

[109] C.G. Bell and A. Newell, *Computer Structures: Readings and Examples*. MacGraw-Hill, 1971.

[110] M. Barbacci, "Instruction Set Processor Specifications (ISPS): The Notation and its Application," *IEEE Transactions on Computers*, vol. C-30, pp. 24–40, Jan. 1981.

[111] D.G. Bradlee and R.E. Henry and S.J. Eggers, "The Marion System for Retargetable Instruction Scheduling," in *Proc. of the Int. Conf. on Programming Language Design and Implementation (PLDI)*, pp. 229–240, 1991.

[112] Y. Bajot and H. Mehrez, "A Macro-Block Based Methodology for ASIP Core Design," in *Proc. of the Int. Conf. on Signal Processing Applications and Technology (ICSPAT)*, Nov. 1999.

[113] F. Engel and J. Nührenberg and G.P. Fettweis, "A Generic Tool Set for Application Specific Processor Architectures," in *Proc. of the Int. Workshop on Hardware/Software Codesign*, Apr. 1999.

[114] M. Gschwind, "Instruction Set Selection for ASIP Design," in *Proc. of the Int. Workshop on Hardware/Software Codesign*, May 1999.

[115] Improv, *Jazz*
*http://www.improvsys.com*, 2001.

[116] S.H. Leibson, "Jazz Joins VLIW Juggernaut – CMP and Java as an HDL Take System-on-Chip Design to Parallel Universe," *Microprocessor Report*, 2000.

[117] Tensilica, *Xtensa*
*http://www.tensilica.com*, 2001.

[118] R. Gonzales, "Xtensa: A configurable and extensible processor," *IEEE Micro*, vol. 20, pp. 60–70, Mar. 2000.

[119] A. Hoffmann and A. Nohl and G. Braun and H. Meyr, "A Survey on Modeling Issues Using the Machine Description Language LISA," in *Proc. of the Int. Conf. on Acoustics, Speech and Signal Processing (ICASSP)*, May 2001.

[120] S. Pees and A. Hoffmann, "Retargetable Simulation and LISA Description of the TMS320C6x DSP," Final Report, Integrated Signal Processing Systems, RWTH Aachen, 1999.

[121] A. Hoffmann and A. Nohl and G. Braun and H. Meyr, "Modeling and Simulation Issues of Programmable Architectures," in *Proc. of the Workshop on Software and Compilers for Embedded Systems (SCOPES)*, Mar. 2001.

[122] S. Pees and A. Hoffmann and H. Meyr, "Retargetable Compiled Simulation of Embedded Processors Using a Machine Description Language," *ACM Transactions on Design Automation for Electronic Systems*, 1999.

[123] S. Pees and A. Hoffmann and H. Meyr, "Retargeting of Compiled Simulators for Digital Signal Processors Using a Machine Description Language," in *Proc. of the Conference on Design, Automation & Test in Europe (DATE)*, Mar. 2000.

[124] T. Gloekler and S. Bitterlich and H. Meyr, "ICORE: A Low-Power Application Specific Instruction Set Processor for DVB-T Acquisition and Tracking," in *Proc. of the ASIC/SOC conference*, Sep. 2000.

[125] A. Nohl, "Development of a Generic Integer C++ Data Type for the use within the Machine Description Language LISA." Advisor: A. Hoffmann, Feb. 1999.

[126] J. Hennessy and D. Patterson, *Computer Architecture: A Quantitative Approach*. Morgan Kaufmann Publishers Inc., 1996. Second Edition.

[127] A. Hoffmann and T. Kogel and A. Nohl and G. Braun and O. Schliebusch and A. Wieferink and H. Meyr, *LISA Manual*. Institute for Integrated Signal Processing Systems (ISS), RWTH Aachen, 2001.

[128] Mentor Graphics, *Seamless*
*http://www.mentor.com/seamless*, 2001.

[129] Cadence, *Cierto*
*http://www.cadence.com/technology/hwsw*, 2001.

[130] Synopsys, *Eaglei*
*http://www.synopsys.com/products/hwsw*, 2001.

[131] Synopsys, *Physical Compiler*
*http://www.synopsys.com/products/unified_synthesis/unified_synthesis.html*, 2001.

[132] Cadence, *Virtuoso Placer*
*http://www.cadence.com/products/vplacer.html*, 2001.

[133] Cadence, *Virtuoso Router*
*http://www.cadence.com/products/vrouter.html*, 2001.

[134] O. Weiss and M. Gansen and T.G. Noll, "A flexible Datapath Generator for Physical Oriented Design," in *Proc. of the European Solid-State Circuits Conference (ESSCIRC)*, Sep. 2001.

[135] A. S. et.al., "Accelerating Concurrent Hardware Design with Behavioral Modeling and System Simulation," in *Proc. of the Design Automation Conference (DAC)*, 1995.

[136] Synopsys, *Vss*
*http://www.synopsys.com/products/vss*, 2001.

[137] Synopsys, *Cyclone*
*http://www.synopsys.com/products/cyclone*, 2001.

[138] Cadence, *Leapfrog*
*http://www.cadence.com/technology/leapfrog*, 2001.

[139] Mentor Graphics, *ModelSim*
*http://www.mentor.com/modelsim*, 2001.

[140] K. Keutzer and S. Malik and A.R. Newton and J.M. Rabaey and A. Sangiovanni-Vincentelli, "System-Level Design: Orthogonalization of Concerns and Platform-Based Design," *IEEE Transactions on Computer-Aided Design*, vol. 19, pp. 1523–1543, Dec. 2000.

[141] Synopsys, *COSSAP*
*http://www.synopsys.com/products/cossap.html*, 2001.

[142] OPNET, *http://www.opnet.com*, 2001.

[143] J. Volder, *"The CORDIC trigonometric computing technique"*. IRE Transactions on Electronic Computers, Sep. 1959. Volume EC-8, no.3, 330-334.

[144] ARC Cores Ltd., *ARCtangent Processor*
*http://www.arccores.com*, 2001.

[145] A. Hoffmann and O. Schliebusch and A. Nohl and G. Braun and O. Wahlen and H. Meyr, "A Methodology for the Design of Application Specific Instruction-Set Processors Using the Machine Description Language LISA," in *Proc. of the Int. Conf. on Computer Aided Design (ICCAD)*, Nov. 2001.

[146] O. Schliebusch and A. Hoffmann and A. Nohl and G. Braun and H. Meyr, "Architecture Implementation Using the Machine Description Language LISA," in *Proc. of the Asia South Pacific Design Automation Conference (ASPDAC)*, Jan. 2002.

[147] O. Schliebusch, "Automatic Synthesis of Pipeline Structures and Instruction Decoder from Abstract Machine Descriptions," Master's thesis, Integrated Signal Processing Systems, RWTH Aachen, Nov. 2000. Advisor: A. Hoffmann.

[148] Nokia, *Multimedia Terminals*
*http://www.nokia.com/multimedia*, 2001.

[149] W. Webb, "Outfitting the Small System Toolbox," *EDN Magazine*, pp. 51–58, Aug. 2001.

[150] R. Cravotta, "Software Development Tools are Growing Up," *EDN Magazine*, pp. 65–70, Jul. 2001.

[151] Motorola, *SC110 DSP Core Reference Manual*, Apr. 2001.

[152] Analog Devices Inc., *TigerSHARC Reference Manual*, Dec. 2001.

[153] R. Leupers and P. Marwedel, *Retargetable Compiler Technology for Embedded Systems – Tools and Applications*. Kluwer Academic Publishers, 2001.

[154] M. Willems and V. Živojnović, "DSP-Compiler: Product Quality for Control-Dominated Applications?," in *Proc. of the Int. Conf. on Signal Processing Applications and Technology (ICSPAT)*, Oct. 1996.

[155] M. Coors and O. Wahlen and H. Keding and O. Lüthje and H. Meyr, "C62x Compiler Benchmarking and Performance Coding Techniques," in *Proc. of the Int. Conf. on Signal Processing Applications and Technology (ICSPAT)*, Nov. 1999.

[156] O. Wahlen and T. Gloekler and A. Nohl and A. Hoffmann and R. Leupers and H. Meyr, "Application Specific Compiler/Architecture Codesign: A Case Study," in *In Proc. of the Conference on Languages, Compilers, and Tools for Embedded Systems (LCTES)*, Jun. 2002.

[157] R. Cravotta, "DSP Directory 2001," *EDN Magazine*, Mar. 2001.

[158] A. Nohl, "Investigations of the Retargetability of Development Tools for Embedded Processors," Master's thesis, Integrated Signal Processing Systems, RWTH Aachen, May 2000. Advisor: A. Hoffmann.

[159] Gintaras R. Circys, *Understanding and using COFF.* O'Reilly and Associate, Inc., first ed., November 1998.

[160] Texas Instruments, *TMS320C6000 Assembly Language Tools - User's Guide*, Nov. 1995.

[161] C. Mills, S. Ahalt, and J. Fowler, "Compiled instruction set simulation," *Software Practice and Experience*, vol. 21, pp. 877–889, Aug. 1991.

[162] S. Pees and V. Živojnović and A. Ropers and H. Meyr, "Fast Simulation of the TI TMS 320C54x DSP," in *Proc. of the Int. Conf. on Signal Processing Applications and Technology (ICSPAT)*, Sep. 1997.

[163] G. Braun and A. Hoffmann and A. Nohl and H. Meyr, "Using Static Scheduling Techniques for the Retargeting of High Speed, Compiled Simulators for Embedded Processors from an Abstract Machine Description," in *Proc. of the Int. Symposium on System Synthesis (ISSS)*, Oct. 2001.

[164] Analog Devices Inc., *VisualDSP++, v2.0*, Apr. 2001.

[165] Texas Instruments, *Code Composer Studio*, May 2001.

[166] Advanced Risc Machines Ltd., *ARM Developer Suite (ADS), v1.0*, Dec. 2000.

[167] A. Hoffmann and S. Pees and H. Meyr, "A Retargetable Tool-Suite for Exploration of Programmable Architectures in SOC-Design," in *Proc. of the Int. Conf. on Signal Processing Applications and Technology (ICSPAT)*, Nov. 1999.

[168] Texas Instruments, *TMS320C54x CPU and Instruction Set Reference Guide*, Oct. 1996.

[169] G. Braun, "Examinations of the Applicability of Compiled Techniques in Retargetable Processor Simulation," Master's thesis, Integrated Signal Processing Systems, RWTH Aachen, Aug. 2000. Advisor: A. Hoffmann.

[170] R. Sites et "*al.*, "Binary Translation," *Comm. of the ACM*, vol. 36, pp. 69–81, Feb. 1993.

[171] GNU – Free Software Foundation, *Debugging with GDB* *http://www.gnu.org/manual/gdb-4.17/html_mono/gdb.html*, 2001.

[172] Technical University of Braunschweig, *DDD - Data Display Debugger* *http://www.gnu.org/softech/ddd*, 2001.

[173] Analog Devices Inc., *ADSP2101 User's Manual*, Sep. 1993.

[174] V. Živojnović and J. Martinez and C. Schläger and H. Meyr, "DSPstone: A DSP-Oriented Benchmarking Methodology," in *Proc. of the Int. Conf. on Signal Processing Applications and Technology (ICSPAT)*, Oct. 1994.

[175] Microsoft, *http://www.microsoft.com*, 2001.

[176] B. Kernighan and D. Ritchie, *The C Programming Language*. Prentice Hall Software Series, 1988.

[177] International Telecommunication Union (ITU), *http://www.itu.int*, 2001.

[178] European Telecommunication Standardization Institute (ETSI), *http://www.etsi.org*, 2001.

[179] The Flow Control Consortium, *Quantum Flow Control Specification, Rev.2.0* *http://www.qfc.org*, 1995.

[180] A. Hoffmann and A. Nohl and G. Braun and H. Meyr, "Generating Production Quality Software Development Tools Using A Machine Description Language," in *Proc. of the Conference on Design, Automation & Test in Europe (DATE)*, Mar. 2001.

[181] LISA Homepage, *http://www.iss.rwth-aachen.de/lisa*. Institute for Integrated Signal Processing Systems (ISS), RWTH Aachen, 2001.

[182] R. Camposano and J. Wilberg, "Embedded System Design," *ACM Transactions on Design Automation for Electronic Systems*, vol. 10, no. 1, pp. 5–50, 1996.

[183] The Open SystemC Initiative (OSCI), *SystemC* *http://www.systemc.org*, 2001.

[184] The Open SystemC Initiative (OSCI), *Functional Specification for SystemC 2.0* *http://www.systemc.org*, 2001.

[185] A. Hoffmann and T. Kogel and H. Meyr, "A Framework for Fast Hardware-Software Co-simulation," in *Proc. of the Conference on Design, Automation & Test in Europe (DATE)*, Mar. 2001.

[186] Synopsys, *CoCentric SystemC Compiler* *http://www.synopsys.com/products/cocentric_systemC/cocentric_systemC.html*, 2001.

[187] GRACE++ Homepage, *http://www.iss.rwth-aachen.de/grace*. Institute for Integrated Signal Processing Systems (ISS), RWTH Aachen, 2001.

[188] R. Camposano, "Design Planing for .25$\mu m$ and beyond," Nov. 1997. presented at 3rd Annual European Symposium: "Towards System on Silicon", Synopsys, Aachen.

[189] A. Ropers, "Techniques for Compiled HW/SW Cosimulation," in *Proc. of the DSP Germany Conference*, 1998.

[190] M. Jung, "Intelligent Generation of Testpattern from an Abstract Machine Description Language," Master's thesis, Integrated Signal Processing Systems, RWTH Aachen, Sep. 2001. Advisor: A. Hoffmann.

[191] F. Fiedler, "SystemC based Data-Path Synthesis using the LEON Sparc 8 Architecture.," Master's thesis, Integrated Signal Processing Systems, RWTH Aachen, Dec. 2001. Advisor: O. Schliebusch and A. Hoffmann.

[192] D. Seal, *ARM –Architecture Reference Manual.* Addison Wesley, 2nd Edition, 2000.

# About the Authors

*Andreas Hoffmann* received his Diploma degree in Electrical Engineering and Information Technology from Ruhr-Universität-Bochum in 1997 with honors. In 1997 he joined the Institute for Integrated Signal Processing Systems (ISS) at Aachen University of Technology (RWTH Aachen) as a research assistent where he was appointed as a chief engineer in 2000. In 2002 he received his PhD degree from RWTH Aachen with honors. He authored numerous technical articles on retargetable software tools and embedded processor design, and received the Best Paper Award at DAC 2002. Since 2002 he is with LISATek GmbH in Aachen, Germany, a spin-off from RWTH Aachen commercializing the LISA based technology, where he heads the engineering department.

Contact information:
Andreas Hoffmann
LISATek GmbH
Technologiezentrum am Europaplatz
Dennewartstrasse 25-27
52068 Aachen, Germany
*email:* andreas.hoffmann@lisatek.com
*web:* http://www.lisatek.com

*Rainer Leupers* received the Diploma and Ph.D. degrees in Computer Science with honors from the University of Dortmund in 1992 and 1997, respectively. From 1993 to 2001 he was a member of the Embedded Systems research group at Dortmund. His research and teaching activities revolve around software development tools for embedded systems, with emphasis on efficient compilers. In 2002, Dr. Leupers joined the ISS institute at Aachen University of Technology as a professor for Software for Systems on Silicon. He authored three books and numerous technical articles on compilers for embedded systems, and he received Best Paper Awards at DATE 2000 and DAC 2002. Additionally, he has been heading the embedded software tools group at ICD (Dortmund) and he is a co-founder of LISATek.

Contact information:
Prof. Rainer Leupers
Software for Systems on Silicon
RWTH Aachen
Templergraben 55
52056 Aachen, Germany
*email:* rainer.leupers@iss.rwth-aachen.de
*web:* http://www.iss.rwth-aachen.de

*Heinrich Meyr* received his M.S. and Ph.D. from ETH Zurich, Switzerland, and spent 12 years in research and management positions in industry. In 1977, he accepted a professorship in Electrical Engineering at Aachen University of Technology (RWTH Aachen) where he heads ISS, the Institute for Integrated Signal Processing Systems, a research institute involved in the analysis and design of complex signal processing systems for communication applications.

As well as being a Fellow of the IEEE, Dr. Meyr has served as Vice President for International Affairs of the IEEE Communications Society. He has published numerous IEEE papers and holds many patents. Dr. Meyr has authored two books, Synchronization in Digital Communications, Wiley 1990 (together with Dr. G. Ascheid), and Digital Communication Receivers, Synchronization, Channel Estimation, and Signal Processing, Wiley 1997 (together with Dr. M. Moeneclaey and Dr. S. Fechtel).

In 1998, Prof. Meyr was a visiting scholar at the UC Berkeley Wireless Research Center (BWRC). During the spring term 2000, the UC Berkeley EE department elected him the "McKay distinguished lecturer" for his series entitled, "Design and Implementation of Advanced Digital Receivers for Wireless Communications." Prof. Meyr was also the 2000 recipient of the prestigious "Mannesmann Innovation Prize" for outstanding contributions to the area of wireless communication. He is a co-founder of LISATek.

Contact information:
Prof. Heinrich Meyr
Integrated Signal Processing Systems
RWTH Aachen
Templergraben 55
52056 Aachen, Germany
*email:* heinrich.meyr@iss.rwth-aachen.de
*web:* http://www.iss.rwth-aachen.de

# Index